SAP ABAP Command Reference

by

Dennis Barrett

ISBN: 0-7596-5912-5

This book is printed on acid free paper.

1stBooks - rev. 01/03/02

Table of Contents

Acknowledgments – First Edition

I first published this book at R/3 release 3.0 and received lots of help from the publisher. That publisher decided to let it go out of print and not to publish an update. According to my contract with them I can't mention their name, but I thank them for their help.

Thanks also to Tipton Cole for every sort of assistance and permission to use the York-Mills Notation, and to John Keating and Hart Graphics for their support and access to the Release 2.2 system.

Acknowledgments – Second Edition

Quite a few people sent me comments and corrections, and several colleagues told or wrote me about tricks and tips for getting more value out of ABAP. Thanks to all of them.

Thanks to SAP America (my employer now) for creating and distributing this fascinating, powerful and rich program, and for access to the 4.x systems where I was able to explore the new features of the language.

Finally, thanks to 1stBooks for providing this creative new way to publish.

Introduction

This book is a reference guide to ABAP, the programming language for SAP. This chapter outlines the structure of ABAP Command Reference. SAP AG, the parent company manufactures several software products: R/2, R/3, CRM, SCM, APO, BW, and others. Most of those products use and are (mostly) written in the ABAP language. For years the most popular product (and the only one using ABAP) was R/3, so we often referred to ABAP as the R/3 language. It's now used in several new products, so I refer in this book to "SAP" running the code, meaning any of the SAP products that use ABAP.

The SAP (we say S. A. P. not "sap") products use the ABAP programming language to construct many of their features, and developers may use ABAP to prepare custom reports, import legacy data, design on-line transactions, and perform other programming tasks. The name stands for "Advanced Business Application Programming."

This reference is intended to provide a concise, accurate and accessible summary of the main elements of SAP and ABAP that a developer needs to do that work, specifically the syntax of commands, some of their options ("additions" in SAP lingo), pointers to related commands, and some of the "magic words" found in the R/3 world.

This book is intended for the convenience of trained, experienced ABAP programmers, and not as an introduction to ABAP or as a teaching tool. However, it may be useful to students of ABAP as a supplement to textbooks and other training materials.

Some of the commands have variants not described herein. In the spirit of a quick reference, I have covered the variants that appear to be in common use and left out the really obscure ones. See the On-line Help for more variants to those commands. I haven't included commands shown in the on-line help as "for internal use only" or as obsolete.

There appear to be some variants which SAP has not documented. Have a look at some native SAP programs, and you'll often find very interesting uses of commands that don't show up in the On-line help or the manuals. I've included very few of these, since they're likely to be unstable between releases, and have labeled them as "undocumented".

The ABAP language grows and changes through the revisions. Some commands and some additions apply only to certain releases. I've identified those where I could.

This reference uses typefaces and symbols as follows:

KEYWORD (ABAP command, event, option, etc.)

SY-VARIABLE - system variables

itab - the literal name of an internal table

f1, f2, f3...- literal field names & formal parameters

a1, a2...- actual parameters

dbtab - literal names of database tables

<value> - either a literal number or string, or a variable

arrayname - literal name of a data structure

[command option] - an option on the main command is enclosed in square braces; the braces are not typed in the program

[command option...] - the option may be repeated some number of times; the braces and ellipsis are not typed in the program

{alternate1 | alternate2} - alternative options are shown separated by pipes (vertical lines) and enclosed in braces; the pipes and braces are not typed in the program

For example, the ASSIGN command and its options (additions)

```
ASSIGN
    { [LOCAL COPY OF] {f1[+p1][(w1)] | (f2) }
    | COMPONENT f3 OF STRUCTURE array1}
    TO <fs> [TYPE t1] [DECIMALS d1].
```

are interpreted as follows:

1. **ASSIGN** must assign to <fs> either
   f1 or
   (f2) or
   COMPONENT f3 **OF STRUCTURE** array1.
2. If it was f1, then the command could actually assign the substring of f1 defined by its offset +p1 and/or its width (w1).
3. If it was either f1 or (f2) and the statement was in a subroutine, then it could assign a **LOCAL COPY** of f1 or (f2) in the subroutine.
4. The command may assign a **TYPE** to the target, and
5. It may assign a **DECIMALS** value to any Type P target.

SAP R/3 provides huge capability to interactive users who enter commands at the menus and screens. In order to have a convenient and concise way to record those commands, I have included the separately-developed **York-Mills Notation** in *Appendix I* with permission of the authors. That specification provides a way to publish the interactive command sequences shown in this reference.

Why Use ABAP?

People use ABAP because it is the language provided with SAP products, and is essentially the only way to develop custom reports, interfaces with other systems, and user transactions for SAP.

Integration in SAP SYSTEMS

Everything in SAP is stored in tables, including the source for ABAP programs. That source can include code and several objects stored separately from the code: "text elements" - labels, headings and the like, graphical screens, and menus. The programming environment is quite visual, and it's difficult to imagine or simulate outside an SAP system.

Rapid Development

Many programming projects are high level rather than low level. That means that they tend not to involve bit-level manipulations or direct operating system calls. Instead, they focus on reading from and writing to tables and files, reformatting the output, and printing reports. With ABAP, the programmer does not need to get involved in the details of how file handles and buffers are manipulated, how memory is allocated, and so on. It's almost understandable even without knowing any ABAP, especially if you are familiar with languages like Basic or Pascal.

ABAP can feel clunky. It appears to have grown out of a blend of assembler, COBOL, RPG and SQL. ABAP commands may look and feel a bit different from each other, depending on their heritage, which leads to the great value of a reference work: it's tough to memorize this command set. The language is powerful for its purpose; you can rather quickly code, test and deliver to the user a program or custom report.

Flexibility

ABAP was not designed in the abstract. It was written to develop the SAP enterprise application and it evolved to serve an ever widening set of business problems that SAP addresses.

The Structure of This Book

This book contains an introductory section, an alphabetical reference to the keywords in the ABAP language and a set of appendices that show detailed information in tables and examples.

Since all the keywords are covered in the alphabetical reference section, an index and glossary would be redundant, and are not included.

About me

I am an applications consultant and project manager for SAP America, assigned to the New York City office. I've been programming and developing database applications over 16 years in several programming languages, and received my certification in ABAP from the SAP Partners Academy in 1996. I graduated from the University of Texas at Austin Mathematics and Physics, and have attended SAP courses in ABAP and in several applications areas. I founded and chaired the Central Texas Chapter of the SAP Americas Users Group (ASUG) and founded the Small and Medium-sized Enterprise Interest Group of ASUG.

My web page address is www.mindspring.net/~dennis.barrett. I'll maintain there a list of corrections to errors found in this book. If you find such errors, please send them to Dennis.Barrett@Bright-Star.net. While I am employed by SAP America, this book was created independently of that employment, and is not an official publication of my employer. Opinions expressed in this book are mine alone, and any errors in this book are mine and not the fault of SAP America.

Change History

Date	Description
September 1997	First Edition published (SAP R/3 Releases 2.2 – 3.1)
December 2000	Out of print
December 2001	Second Edition published by 1stBooks.com (through SAP releases 4.6)

ABAP Overview

This book is designed as a reference guide for the ABAP language, rather than an introductory text. However, some aspects of the language are better summarized in a short narrative rather than in a bullet list in the reference section. Therefore, this chapter puts the reference material in context by giving an overview of the ABAP language.

Running ABAP

The simplest way to run an ABAP program (frequently called "an ABAP") from inside SAP R/3 is to execute the transaction SA38 (type /NSA38 <Enter> in the command field), then type the name of the ABAP in the resulting blank field, and <Enter>. There is no way to run an ABAP outside of an SAP system.

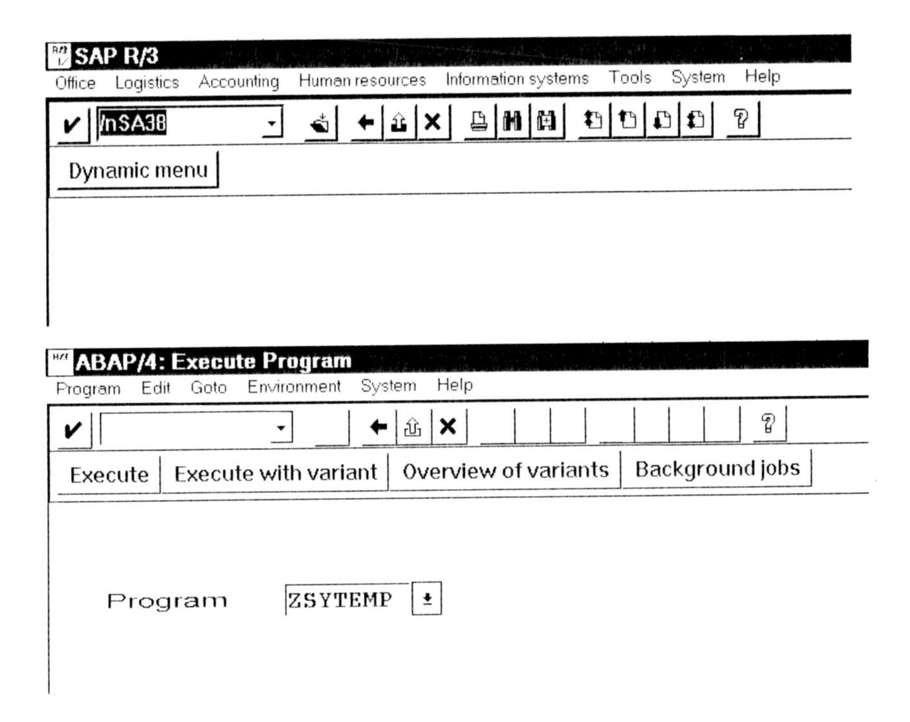

An ABAP Script

An ABAP program consists of an internal SAP file containing a series of ABAP commands. Commands are written in what looks like an amalgam of assembler, COBOL, RPG and SQL. In fact, that's pretty much what it is. The internal file can be uploaded from and downloaded to a normal ASCII text file for documentation or other purposes, but only the internal file can be executed.

The easiest way go into the ABAP Editor to create or edit programs is by “going into SE38”, that is, typing /nSE38 <Enter> in the command field.

ABAP code can be quite free-flowing. The broad syntactic rules governing where a statement starts and ends are:

- Leading white space is ignored. You can start an ABAP statement anywhere you want, at the beginning of the line, indented for clarity (recommended), or even right-justified (definitely frowned on) if you like. Lines may be up to 72 characters long.
- Statements are terminated with a period.
- White space outside of string literals is irrelevant; one space is as good as a hundred. That means you can split most statements over several lines for clarity, and you can place several short statements on the same line.
- Literal strings are delimited by single-quotes (').
- Anything on a line after a double-quote (“) is ignored, and any line beginning with an asterisk (*) is ignored. Use this to pepper your code with useful comments. The next line break ends the comment.
- ABAP statements take the form:

  ```
  Keyword parameters data_objects period.
  ```

  Many statements may follow the keyword with a colon (:) and accept comma-separated multiple predicates; each element (keyword, parameter, grouping parentheses) must be separated by one or more spaces or a line break. Case is not distinguished by the processor, except for string comparisons; you are free to use case to enhance readability. A statement is terminated by a period.

Here’s an ABAP statement illustrating the simple statement:

```
    WRITE 'This book has been available for 4 years.'.
```

No prizes for guessing what happens when ABAP runs this code. It prints

```
    This book has been available for 4 years.
```

Printing more text is a matter of either stringing together statements or giving multiple arguments to the print function:

```
WRITE 'This book has been available for 4 years'.
WRITE: / 'It was originally called',
         '''SAP R/3 ABAP/4 Command Reference''',',
         /, 'because it focused on R/3,',
```

```
        'and ABAP was called ABAP/4 then.'.
```

If the "/" doesn't look familiar, don't worry; it simply prints a newline character. In other words, the report will go to the start of the next line. In ABAP you generally place the newline before the string instead of after it. The first slash needn't be comma-separated, but all the rest of them must be.

I strung two single-quote characters in a row to force a quote character in the string. See how it prints out below.

Notice that I strung together the second sentence in one WRITE statement using commas. Also see that the second sentence spans four lines, and the extra spaces on the third, fourth and fifth statement lines don't show up in the print lines. ABAP does however insert one space between each field that it WRITEs, so I needn't include spaces inside the strings.

```
This book has been available for 4 years.
It  was  originally  called  'SAP  R/3  ABAP/4  Command  Reference',
because it focused on R/3, and ABAP was called ABAP/4 then.
```

That's not at all typical of a ABAP program though; it's just a linear sequence of commands with no structural complexity. The "Branch Control" section later in this overview introduces some of the constructs that make ABAP what it is. For now, we'll stick to simple examples like the preceding for the sake of clarity.

Data Types

ABAP has several native data types, and has the means for you to define your own complex types from the native ones. Those complex types can include arrays whose elements have differing types; indeed any one such element can be a table with any number of records. See "Type" in the alphabetical section for a description of each of the native types.

Variables

Variables (often called "fields" in SAP lingo) may be of any native or user-defined type. Variable names are limited to 30 characters and must include at least one non-number character; they may not include spaces, parentheses, plus (+), minus or hyphen (-), commas or periods. These names are case-insensitive.

ABAP converts automatically between types where it makes sense. All of these conversions are handled implicitly, leaving the programmer free to concentrate on what needs to be done rather than the low-level details of how it is to be done.

We can use variables to develop the earlier example script using some string variables:

```
DATA:

    what(12) VALUE 'This book',    "defaults to character type
    howlong TYPE I VALUE '4',
    called(24) VALUE 'was originally',
    name(50) VALUE 'SAP R/3 ABAP/4 Command Reference'.

    WRITE: what, 'has been available for', howlong, 'years.',
          /, 'It', called, 'called ''', name NO-GAP,
          ''',' NO-GAP,
          /, 'because it focused on R/3,',
          'and ABAP was called ABAP/4 then.'
```

```
This book has been available for 4 years.
It was originally called 'SAP R/3 ABAP/4 Command Reference',
because it focused on R/3, and ABAP was called ABAP/4 then.
```

The three single quotes open or close a string while forcing a quote character at the beginning or end of the string.

Branch Control

The examples we've seen so far have been quite simple, with little or no logical structure beyond a linear sequence of steps. ABAP has the following branch control mechanisms; see their descriptions in the alphabetical section for details.

```
IF condition...ELSEIF condition2...ELSE...ENDIF
DO [n TIMES]...ENDDO
WHILE condition...ENDWHILE
```

and for specific situations:

```
LOOP...ENDLOOP
```

Conditional Expressions

You may use several forms of conditional (logical) expressions. See "Condition" and "Operators" in the alphabetical section for the details.

Subroutines

Subroutines in ABAP are code blocks whose definitions begin with `FORM formname` and end with `ENDFORM`. You can pass arguments to and from forms by value or by reference, and you can pass working tables to and from them. A form is called by the `PERFORM formname` command. Forms may call other forms and they may call themselves (recursion).

Alphabetical Reference

= *(equals sign)*

Description

"Equality" relational operator (same as `EQ`), and assignment operator. Multiple assignments are supported, and are processed from right to left.

Example

```
x = y = 3.
WRITE: 'x =', x, '; y =', y.
```

prints `x = 3 ; y = 3`

See also *Operators*, `MOVE`, `WRITE TO`

[]

Syntax

`itab[].`

Description

itab[] always refers to the internal table itself, not to its header line.

See also *itab*, *Work Area*

*

Syntax

`*tablename`

Description

Alternative work area for the table `tablename`, declared in the `TABLES` statement.

See also `TABLES`, *Work Area*

ABS

Syntax

`ABS``( x ).`

Description

Returns the absolute value of any number `x`.

See also *Arithmetic functions*

ACOS

Syntax

```
ACOS( y ).
```

Description

Returns the arc-cosine of floating point number y, for y between -1 and +1, and **ACOS**(y) between 0 and π;

See also *Arithmetic functions*

ADD

Syntax

```
ADD a TO b.
```

Description

Equivalent to `b = b + a.` Non-numeric fields are converted. See "Type Conversions" for conversion information.

See also DIVIDE, MULTIPLY, SUBTRACT

ADD *(series)*

Syntax

```
ADD a_1 THEN a_2 UNTIL a_n TO b.
ADD a_1 THEN a_2 UNTIL a_n [ACCORDING TO rtab] GIVING b.
ADD a_1 FROM indx_1 TO indx_n GIVING b.
```

Description

ADD...THEN adds together all the elements a_i in the series (in a structure) where each element is the same type and length. **TO** adds the sum to the current contents of b, and **GIVING** overwrites its contents. **ACCORDING TO** rtab includes only the elements whose indices are specified in rtab which was created in a RANGES or SELECT-OPTIONS command. **ADD...FROM** adds together those elements whose indices start with indx_1 and end with indx_n of the series (in a structure) that begins with a_1. The elements must have the same type and length.

See also RANGES, SELECT-OPTIONS

ADD-CORRESPONDING

Syntax

```
ADD-CORRESPONDING array1 TO array2.
```

Description

If array1 & array2 are structured work areas such as header lines, then this command adds like-named fields in array1 and array2.

Example

```
ADD array1-key1 TO array2-key1.
ADD array1-key2 TO array2-key2.
etc.
```

See also DIVIDE-CORRESPONDING, MULTIPLY-CORRESPONDING, SUBTRACT-CORRESPONDING

Addition

Description

An “addition” in SAP lingo is an optional parameter which expands the function of its base command.

Example

In the description of **APPEND** below, the six additions are

- wa **TO** itab1
- **INITIAL LINE TO** itab1
- **LINES OF** itab2 **TO** itab1
- **FROM** ndx1
- **TO** ndx2
- **SORTED BY** f1.

In this reference manual, optional parameters are indicated by the square braces [ ] in the command description, and alternatives among those options are separated by the “pipe” or vertical bar |. If one of several parameters must be used, the list of parameters is enclosed in curly braces {}, and the parameters in the list are separated by vertical bars. Don’t type the braces or vertical bars in your programs.

ADJACENT DUPLICATES

See DELETE ADJACENT DUPLICATES

ALE

Description

“Application Linking and Enabling” – the protocol for communicating between any number of SAP systems, normally all operated by one enterprise. ALE uses IDOCS as the data container.

See also EDI, IDOCS

ALIASES

Description

Defines alias names for an interface component in ABAP Objects. This book doesn’t discuss ABAP Objects programming.

See also *Object-oriented programming*

APPEND

Syntax

```
APPEND
[wa
| INITIAL LINE
| LINES OF itab2 [FROM ndx1] [TO ndx2]]
TO itab1
[SORTED BY f1].
```

Description

Appends a new record to the end of `itab1` from its header line, from work area `wa`, or from an initial-value (that is, cleared) structure, or appends records from `itab2` to the end of `itab1`, starting with the first record of `itab2`, or from record `ndx1` if it is specified, and continuing to the last record, or record `ndx2` if it is specified. There's no limit to the number of records `itab1` may accept. `wa` & `itab2` must have the same structure as `itab1`.

SORTED BY `f1` sorts `itab1` on field `f1` after the record is appended, then truncates the table to the number of records in the **OCCURS** declaration. (This option is only documented for the header line and work area options.)

`SY-TABIX` contains the single record number of the new entry in `itab1`, or in the case of **LINES OF** `itab2`, the record number of the last record entered.

See also COLLECT, DELETE, INSERT, LOOP, MODIFY, READ, SELECT, SORT, WRITE TO

Application Server

Description

The "middle layer", where every SAP product's functional code resides and runs (the application programs, ABAP editor, etc.). It is below the SAPGUI Presentation Client, and above the Database Server. The Application server is also an instance of the SAP system.

See also *Database Server, Instance, Presentation Server, SAPGUI*

Arithmetic functions

Arguments are: `( x )` all numbers, `( y )` floating point numbers, and `( s )` strings. Notice that the arguments must be separated from the parentheses by spaces.

ABS `( x ).`	Absolute value of x
ACOS `( y ).`	Arc-cosine of y, between 0 and π
ASIN `( y ).`	Arc-sine of y, between -π/2 and π/2
ATAN `( y ).`	Arc-tangent of y, between -π/2 and π/2
CEIL `( x ).`	Ceiling of x, that is, the smallest integer not less than x

COS(y).	Cosine of y, for y in radians
COSH(y).	Hyperbolic Cosine of y
EXP(y).	Exponential of y, that is, **e****y for **e** =2.7182818284590452
FLOOR(x).	Floor of x, that is, the largest integer not greater than x
FRAC(x).	Fractional part of x, that is, the decimal portion of x
Integer(x).	use **TRUNC**(x).
LOG(y).	logarithm base *e* of y, for y > 0
LOG10(y).	logarithm base 10 of y, for y > 0
SIGN(x).	Sign of x: 0 → 0; > 0 → 1; < 0 → -1
SIN(y).	Sine of y, for y in radians
SINH(y).	Hyperbolic Sine of y
SQRT(y).	Square root of y, for y ≥ 0
STRLEN(s).	Number of characters in s, to the last non-blank character
TAN(y).	Tangent of y, for y in radians
TANH(y).	Hyperbolic Tangent of y
TRUNC(x).	Truncated x, that is, the integer portion of x

Arithmetic Operators

See Operators

ASIN

Syntax

ASIN(y).

Description

Returns the arc-sine of floating point number y, for y between -1 and +1, and **ASIN** (y) between -π/2 and π/2.

See also *Arithmetic functions*

ASSIGN

Syntax

ASSIGN
{ **[LOCAL COPY OF]** {f1[+p1][(w1)] | (f2) }
| **COMPONENT** f3 **OF STRUCTURE** array1}
TO <fs> **[TYPE** t1**]** **[DECIMALS** d1**]**.

Description

Points the field symbol <fs> to the field f1, or (4.0 options) the field whose name is stored in the field f2. p1 is a literal or variable offset, w1 is a literal or variable width. If p1 + w1 > the width of f1, then <fs> will point to the undefined bytes beyond the end of f1; to avoid this use "*" for w1. **LOCAL COPY OF** f in a subroutine (FORM) creates a local copy of the global variable f and points the local field symbol to it. **COMPONENT** f3

OF STRUCTURE array1 points the field symbol <fs> to element f3 of array1. If f3 is Type c then it is treated as the name of the element in the array, otherwise it is treated as the index of the desired element in the array. **TYPE** t1 assigns Type t1 to the field symbol at runtime. <fs> defaults to the type of the assigned field otherwise. **DECIMALS** d1 only applies to Type P.

SY-SUBRC is only assigned for the indirect addressing case:

= 0 if the field was successfully assigned

= 4 otherwise

See also *Appendix G.2*, FIELD-SYMBOLS

AT FIELDGROUP1

Syntax

AT fieldgroup1 **[WITH** fieldgroup2]...**ENDAT.**

Description

Condition in a **LOOP-ENDLOOP** on a sorted extracted dataset. True when the current record was created by the **EXTRACT** on fieldgroup1. **WITH** fieldgroup2 is true when the current record was created by the **EXTRACT** on fieldgroup1 and the immediately following record was created by the **EXTRACT** on fieldgroup2.

See also EXTRACT, FIELD-GROUPS

AT END

Syntax

AT END OF f1...**ENDAT.**

Description

Condition in a **LOOP-ENDLOOP** on a sorted itab or a sorted extracted dataset. True when the value of f1 or one of the fields to the left of f1 will change at the next record. Also true at the last record. Within this **AT...ENDAT** code block, all fields in the header line to the right of f1 are filled with '*'. In an extracted dataset, f1 must be in the sort key. Don't use this command if the loop includes a **FROM**, **TO** or **WHERE** statement.

See also CNT, SUM, SUM()

AT FIRST

Syntax

AT FIRST...ENDAT.

Description

Condition in a **LOOP-ENDLOOP** on an itab or an extracted dataset. True during the first iteration. Within this **AT...ENDAT** code block, the header line does <u>not</u> contain table data. Don't use this command if the loop includes a **FROM**, **TO** or **WHERE** statement.

See also SUM

AT LAST

Syntax

AT LAST...ENDAT.

Description

Condition in a **LOOP-ENDLOOP** on an itab or an extracted dataset. True during the last iteration. Within this **AT...ENDAT** code block, the header line does <u>not</u> contain table data. Don't use this command if the loop includes a **FROM**, **TO** or **WHERE** statement.

See also CNT, SUM, SUM()

AT LINE-SELECTION

Syntax

AT LINE-SELECTION.

Description

Event for a selection (F2 or double-click with the cursor on a valid line) in an interactive report. A valid line is one created by statements such as **WRITE**, **ULINE OR SKIP**. Fields stored in the **HIDE** area are updated to the values of the selected line. The following system fields are updated to the current status.

Field Name	Description
Runtime:	
SY-TITLE	Title of report from attributes or text fields
List Generation:	
SY-COLNO	Current column in list
SY-LINCT	Page length in list lines (from REPORT)
SY-LINNO	Current line in list
SY-LINSZ	Page width in columns (from REPORT)
SY-PAGNO	Current page in list
SY-PFKEY	PF-Status of the displayed list
SY-SCOLS	Number of columns in window
SY-SROWS	Number of lines in window
Interactive Reporting:	
SY-CPAGE	Current page number
SY-CUCOL	Cursor position (column) on screen

Field Name	Description
SY-CUROW	Cursor position (line) on screen
SY-LILLI	Number of selected list line
SY-LISEL	Contents of the selected line as a string
SY-LISTI	Index of selected list (0=base, 1=detail #1 etc.)
SY-LSIND	Index of displayed list (0=base, 1=detail #1 etc.)
SY-LSTAT	Status information for each list level
SY-MSGLI	Contents of the message line (line 23)
SY-STACO	Number of first displayed column
SY-STARO	Number of first displayed line on this page
SY-UCOMM	Command field function entry

See also *Events, okcodes*

AT NEW

Syntax

`AT NEW f1...ENDAT.`

Description

Condition in a **LOOP-ENDLOOP** on a sorted itab or a sorted extracted dataset. True when the value of f1 or one of the fields to the left of f1 has just changed. Also true at the first record if the first record value of f1 differs from that of the header line. Within this **AT...ENDAT** code block, all fields in the header line to the right of f1 are filled with '*'. In an extracted dataset, f1 must be in the sort key don't use this command if the loop includes a **FROM, TO** or **WHERE** statement.

See also SUM

AT PFNN

Syntax

`AT PFnn.`

Description

Event triggered for function key nn. Since function key assignments aren't fixed, your program may be more stable if you use AT USER-COMMAND. The HIDE area fields and the system fields are updated as described in AT LINE-SELECTION.

See also *Events*

AT SELECTION-SCREEN

Syntax

```
AT SELECTION-SCREEN
[ ON ps
| ON EXIT COMMAND
| ON END OF s
| ON VALUE REQUEST FOR ps_lmts
| ON HELP REQUEST FOR ps_lmts
| ON RADIOBUTTON GROUP r
| ON BLOCK b
| OUTPUT].
```

Description

Events triggered in the selection screen of the program's logical database, in the order in which their corresponding object-creation statements (that is, **PARAMETERS**, **SELECT-OPTIONS**, **SELECTION-SCREEN**) appear in the code except as mentioned below. Issuing an "E"-type message will return focus to the selection screen and to the offending field. **OUTPUT** is triggered at the PBO, before the selection screen is displayed.

Two events are triggered and processed while in the screen. ps_lmts is either the name of a parameter or the name of a select-option concatenated with "-HIGH" or "-LOW" to identify the subject field. These events can provide specialized choices or help. **ON VALUE REQUEST** is triggered if the user presses F4 (*Possible entries*) while the cursor is in the field or clicks on the button beside the field; SAP processes this block <u>instead</u> of displaying a check table or the Dictionary field values. **ON HELP REQUEST** is triggered if the user presses F1 (*Help*) while the cursor is in the field; SAP processes this block <u>instead</u> of displaying a check table or the Dictionary field documentation.

The remaining events are triggered after the PAI completes. They are processed in the order in which the **PARAMETER**, **SELECT-OPTION** and **SELECTION-SCREEN END OF BLOCK** commands appear in the selection screen definition, followed by the basic event. **ON** ps is triggered if the user completed **PARAMETER** ps or **SELECT-OPTION** ps. **ON END OF** s is triggered if user filled in the ranges for **SELECT-OPTION** s. **ON RADIOBUTTON GROUP** r is triggered if the user selected a radiobutton in group r. **ON BLOCK** b is triggered when all objects defined in block b have been completed. **ON EXIT COMMAND** is triggered if the user selected a command that has been set up as an exit command (typically "Cancel", "Back" and "Exit"). Finally, the basic event **AT SELECTION-SCREEN** is triggered after the others complete.

See also *Events,* MESSAGE, PARAMETERS, SELECT-OPTIONS

AT USER-COMMAND

Syntax

```
AT USER-COMMAND.
```

Description

Event triggered by all user commands defined in the menu painter. The value of the triggered code is available in SY-UCOMM for use in a CASE statement. Several system commands are trapped and processed by SAP, so those commands don't trigger this event; they include the command field entries /...and %..., PICK, PFn, PRI (print), BACK, RW (rollback work, that is, cancel), and the scroll functions P...; the **HIDE** area fields and the system fields are updated as described in **AT LINE-SELECTION**

See also *Events, okcodes*

ATAN

Syntax

```
ATAN( y ).
```

Description

Returns the arc-tangent of floating point number y, for **ATAN**(y) between -$\pi/2$ and $\pi/2$.

See also *Arithmetic functions*

Attributes

See **DESCRIBE...** (fields, itabs and lists)

FORMAT and **LOOP AT SCREEN** (screens)

GET PROPERTY (OLE objects)

MODIFY ... LINE (list line)

AUTHORITY-CHECK

Syntax

```
AUTHORITY-CHECK OBJECT objectname
ID f1 FIELD v1
....
ID fn FIELD vn.
```

Description

Tests whether the user has access to the named authorization object and returns SY-SUBRC = 0 if the user has the authorization. Each authorization check object is defined with up to ten specific identifiers, f1 - fn, referred to as **ID**s. The **AUTHORITY-CHECK** statement must specify each such identifier, and provide each with either a required value v1 - vn or the null test value **DUMMY**. The user must have a matching value to every non-**DUMMY** identifier in the user's authorization profiles to be authorized access to objectname. Those matching values need not all be in the same profile.

Authorization objects and their required identifiers may be viewed in /SU21. The user can view his or her own authorization failure in /SU56. For an authorization test that controls the entire program, you may limit access by filling in /SE38 *Attributes <Authorization Group with a value that matches the profile of the permitted users.

SY-SUBRC = 0 - user is authorized access to `objectname`
= 4 - user is NOT authorized access to `objectname`
= 8 - too many parameters
= 12 - object name not in user master control record
= 16 - no profile in user master control record
= 24 - field names don't match those of `objectname`
> 26 - user master record has incorrect structure

BACK

Syntax

BACK.

Description

In the `TOP-OF-PAGE` code block, **BACK** moves the cursor up to first line of the top-of-page area. In the list generation loop and in the `END-OF-PAGE` code block, **BACK** moves the cursor up to first line of the list (after the `TOP-OF-PAGE` output). If used in an `END-OF-PAGE` called by a `RESERVE` statement, then it moves the cursor to the top of the end-of-page area..

See also `END-OF-PAGE`, `RESERVE`, `TOP-OF-PAGE`

BAPIs

"Business Application Programming Interfaces" are remote function modules (RFCs) specifically designed to provide transaction capability to external programs such as Internet front ends.

Batch Data Communications (BDC)

"BDC session" = "Batch Input session" provides a means for bringing external data from a sequential file into SAP, using a defined transaction. The steps are to declare an `itab LIKE bdcdata` then read a record in the sequential file, populate `itab`, then call a function that walks the transaction screens using the itab contents. `bdcdata` is a data dictionary structure. You must thoroughly understand the transaction before you can code the populating of `itab`. Use `/SHDB` to record the steps in the transaction and `/SM35`, `/SM36` & `/SM37` to set up, review, analyze and process batch jobs. The Command field commands are:

Command	Action
/n	terminate current transaction and mark as incorrect
/bdel	delete current BDC from session
/bend	terminate current BDC session & mark as incorrect
/bda	display all transaction steps
/bde	display errors only

See `CALL TRANSACTION`, *Appendix G.1* for an example of a BDC session.

Batch Jobs

You can schedule programs to run at specific times and frequencies, or when triggered by an event. Use `/SM35` to review or monitor batch jobs and `/SM36` to interactively set up a job, or use the following commands in your program:

```
CALL TRANSACTION job_open.
  ...
CALL TRANSACTION job_submit.
  ...
CALL TRANSACTION job_close.
```

Hint: to insert function module calls into your program, use [Pattern.

Bell

There's no `BELL`, `BEEP` or `SOUND` command documented for ABAP.

BINARY SEARCH

See `READ TABLE itab`

Boolean expressions

See *Conditions, Operators*

BT

Relational operator "BeTween" used in "ranges" table.

See also `RANGES`, `SELECT-OPTIONS`

BREAK-POINT

Syntax

`BREAK-POINT [n].`

Description

Halts the running program at this command and starts debugging mode. **n** gives you an easy way to identify the breakpoint when processing halts. The debugger normally issues a **COMMIT WORK**, so the program state will very likely differ between debug mode and normal running mode. Since **COMMIT WORK** clears table cursors, don't insert a **BREAK-POINT** in a **SELECT...ENDSELECT** loop.

BREAK username

Syntax

`BREAK username.`

Description

Issues a BREAK-POINT if **username** equals SY-UNAME.

Business Application Programming Interface

See *BAPI*

CA

String comparison operator "Contains At least one character from".

See *Operators*

Call a program

To execute an ABAP from your terminal use /SA38, /SE 38 or /SE37. To call another ABAP program or report from within an ABAP see SUBMIT. To launch an operating system (that is, UNIX or Windows NT) program on the application server from within an ABAP see OPEN DATASET...FILTER, and /SM49. To launch a program on the presentation server (local workstation) see WS_EXECUTE.

CALL CUSTOMER-FUNCTION

Syntax

`CALL CUSTOMER-FUNCTION f1.`

Description

Calls a function written by the user in a user exit of an SAP program. f1 may be up to 3 characters long. The function name will be EXIT_ plus the SAP module pool name plus f1.

Example

If `f1 = '001'` and the module pool name is `'SAPMABCD'`, then the function module name is `'EXIT_SAPMABCD_001'`.

See also *User Exits*

CALL DIALOG

Syntax

```
CALL DIALOG dcode
[AND SKIP FIRST SCREEN]
[EXPORTING f1 [FROM g1]
   ...
   fn [FROM gn]]
[IMPORTING f1 [TO g1]
   ...
   fn [TO gn]]
[USING itab MODE mode].
```

Description

Calls the dialog transaction `dcode`. Use this command in place of **CALL TRANSACTION** in batch processes. **EXPORTING** passes the arguments (fields, structures and itabs) to `dcode`. **IMPORTING** returns the arguments (fields, structures and itabs) from `dcode`. **FROM** and **TO** list the calling program's names for those arguments if they differ from `dcode`'s names. See **CALL TRANSACTION** for definitions of the other parameters. To exit from the dialog module use the command **LEAVE PROGRAM**. A called dialog module will not process update requests, therefore the calling program must explicitly or implicitly issue a **COMMIT WORK** after the dialog module completes.

See also `CALL FUNCTION`, `CALL TRANSACTION`, `SUBMIT`

CALL FUNCTION

Syntax

```
CALL FUNCTION fmod [IN {UPDATE | BACKGROUND} TASK]
  [DESTINATION 'remote system name'
   [STARTING NEW TASK f 'task name
    [PERFORMING 'formname' ON END OF TASK]]]
  EXPORTING  f1 = a1
             f2 = 'string'
  IMPORTING  f3 = a3
             f4 = a4
  CHANGING   f5 = a5
  TABLES     tab = itab          (passed by reference)
  EXCEPTIONS e1 = subrc1
             e2 = subrc2
                 ...
```

Description
Invokes a function module. Hint: to insert function module calls into your program, use [Pattern. **IN UPDATE TASK** sends the change requests to the Update Server (typically in the same box as the database) to hand off the work load from the Application Server. **IN BACKGROUND TASK** sets up execution in another work process to start at the next **COMMIT WORK**. Use **DESTINATION** to issue a synchronous remote function call (RFC), and use **STARTING NEW TASK** to make the RFC asynchronous. See the discussion in RECEIVE RESULTS for the use of **PERFORMING...ON END OF TASK**. **EXPORTING** passes arguments (fields, structures & itabs) by name from the calling program to the function module. **IMPORTING** returns arguments back to the calling program from the function module. **CHANGING** passes arguments from the calling program to the function module and returns the changed values to the calling program. **TABLES** points to itabs; the table names must be declared in the function. **EXCEPTIONS** lists the exceptions that the calling program should be prepared to handle, and the return value for each; SY-SUBRC will contain the value assigned to the exception returned.
See also CALL DIALOG, CALL TRANSACTION, RECEIVE RESULTS, SUBMIT, *Work process*

CALL METHOD *(type one)*

Syntax
CALL METHOD OF obj1 method1 [= f1] **[NO FLUSH]**
  [EXPORTING parm1=f2 [parm2=f3...]] .
Description
Calls a method of an external (OLE2) object. Optionally assigns the method's return value to f1. **NO FLUSH** continues OLE2 bundling even if the next statement isn't an OLE2 command. **EXPORTING** passes field values f2... to the method's positional parameters parm1...
See also CREATE OBJECT, FREE OBJECT, GET PROPERTY, SET PROPERTY

CALL METHOD *(type two)*

Description
Calls an ABAP Objects method.
This book doesn't discuss ABAP Objects programming.
See also *Object-oriented programming*

CALL SCREEN

Syntax

```
CALL SCREEN scr
[STARTING AT c1 r1
[ENDING AT c2 r2]].
```

Description

Displays screen `scr`. **`STARTING AT`** displays a modal dialog box with the upper left corner at column c1, row r1. **`ENDING AT`** places the lower right corner of the modal dialog box at column c2, row r2. If **`ENDING AT`** is not specified, the lower right position will be determined by the size of `scr`. Use **`SET SCREEN`** `0`**`. LEAVE SCREEN.`** or **`LEAVE TO SCREEN`** `0` to leave this screen. Table cursors are cleared by **`COMMIT WORK`** which is invoked by **`CALL SCREEN`**, so this command shouldn't be used in a **`SELECT`**... **`ENDSELECT`** loop.

See also `COMMIT WORK`, `LEAVE SCREEN`, `SET SCREEN`, `WINDOW`

CALL SELECTION-SCREEN

Syntax

```
CALL SELECTION-SCREEN scr
[STARTING AT c1 r1
[ENDING AT c2 r2]].
```

Description

Displays the selection screen `scr` that is the program start-up screen or that was created using the `SELECTION-SCREEN [BEGIN | END] OF SCREEN` commands. **`STARTING AT`** displays a modal dialog box with the upper left corner at column c1, row r1. **`ENDING AT`** places the lower right corner of the modal dialog box at column c2, row r2. If **`ENDING AT`** is not specified, the lower right position will be determined by the size of `scr`. Control returns to the calling program when the user selects <Continue> or <Cancel>. If the user selects <Cancel>, then the values of the `SELECT-OPTIONS` and `PARAMETERS` fields are remain as they were before the screen is called.

SY-SUBRC	
0	User selected <Continue>
4	User selected <Cancel>

CALL TRANSACTION

Syntax

```
CALL TRANSACTION tcode [AND SKIP FIRST SCREEN]
 USING bdc_table
 [MODE <display mode>]
 [UPDATE <update mode>].
```

Description

Starts the transaction `tcode` for a Batch Data Communications (BDC) session; **See** *Appendix B* for lists of transaction codes. **`AND SKIP FIRST SCREEN`** skips the entry screen; this works only if all the entry parameters are available as Parameter IDs (PIDs). See `SET PARAMETER` for details.

`MODE` - May be a literal or variable.

- `A` Display all steps (default).
- `E` Display errors only.
- `N` No display.

`UPDATE` - May be a literal or variable.

- `A` Continues processing of the calling program while `tcode` proceeds in the background (default).
- `S` Waits for completion of `tcode` before the calling program continues.

To depart from transaction `tcode`, use the command `LEAVE PROGRAM`.

Note**:** Table cursors are cleared by `COMMIT WORK` which is invoked by `CALL TRANSACTION`, so **`CALL TRANSACTION`** mustn't be used in a `SELECT`... `ENDSELECT` loop.

SY-SUBRC	Description
0	Successful
<1000	Error in dialog program
>1000	Error during batch input processing

System Fields	Description
SY-MSGID	Message ID
SY-MSGTY	Message type (`E`, `I`, `W`, `S`, `A` - see `MESSAGE`)
SY-MSGNO	Message number
SY-MSGV1	Message variable 1
SY-MSGV2	Message variable 2
SY-MSGV3	Message variable 3
SY-MSGV4	Message variable 4

See also *Batch data communications*, `CALL DIALOG`, `CALL FUNCTION`, `MESSAGE`, `SUBMIT`

Case: upper & lower

See TRANSLATE

CASE

Syntax

```
CASE field.
  ...
  WHEN value1.
  ...
  WHEN value2.
  ...
  [WHEN OTHERS].
  ...
ENDCASE.
```

Description

Executes the statements between **CASE** and the first **WHEN**. (These statements may be reported as errors by the extended program checker.) Then it finds the first **WHEN** clause whose `value` equals `field`, and executes the statements between that clause and the next **WHEN** or the **ENDCASE**, whichever it encounters first. Any number of **WHEN** statements are allowed. Typically `field` is a variable; `value..` may be variables or literals. **WHEN OTHERS** is optional, only one is allowed and it must be the last **WHEN** clause. Statements between **WHEN OTHERS** and **ENDCASE** are executed only if no other **WHEN** has a matching value.

See also DO, IF, WHILE

CATCH

Syntax

```
CATCH SYSTEM-EXCEPTIONS <e1> = <r1>...
```

Description

In the **CATCH-ENDCATCH** code block, any "catchable" system error or error class `e1` sends process flow to the command following the **ENDCATCH** and populates SY-SUBRC with the numeric literal value `r1`. See online help for catchable errors and error classes. You may nest **CATCH-ENDCATCH** code blocks indefinitely.

SY-SUBRC		
	0	no errors were caught
	r1	error e1 was caught

CEIL

Syntax

CEIL(x).

Description

Returns the ceiling of any number x, that is, the smallest integer not less than x.

See also *Arithmetic functions*

CHAIN

Description

For use in module pools only. In Flow Control, **CHAIN** groups together several events for common processing.

See *Flow Control,* MODULE

CHECK

Syntax

CHECK {<condition> | f1 | **SELECT-OPTIONS**}.

Description

Option 1. (condition) Processing continues with the next command if <condition> is true. If <condition> is false in a repetitive block (DO, LOOP, SELECT, WHILE) then proccessing jumps to the end of the block for the next iteration. If <condition> is false in a non-repetitive block (AT..., FORM, FUNCTION, MODULE) then it terminates processing of the block.

Option 2. (f1) In a **GET** event, processing continues if the field f1 corresponds to its **SELECT-OPTIONS** value, otherwise the event is terminated and subordinate tables are not processed.

Option 3. (**SELECT-OPTIONS**) In a **GET** event, processing continues if all the fields correspond to their **SELECT-OPTIONS** values. Otherwise the event is terminated and subordinate tables are not processed. See *Appendix G.3* for an example of logical database processing (**GET** events).

See also *Condition*, CONTINUE, *Events,* EXIT, LEAVE, REJECT, STOP

CLASS

Description

Declares an ABAP Object Class.

This book doesn't discuss ABAP Object programming.

See also *Object-oriented programming*

CLASS-DATA

Description

Declares static attributes of an ABAP Objects class or interface.

This book doesn't discuss ABAP Objects programming.

See also *Object-oriented programming*

CLASS-EVENTS

Description

Declares static events of an ABAP Objects class or interface.

This book doesn't discuss ABAP Objects programming.

See also *Object-oriented programming*

CLASS-METHODS

Description

Declares static methods of an ABAP Objects class or interface.

This book doesn't discuss ABAP Objects programming.

See also *Object-oriented programming*

CLEAR

Syntax

```
CLEAR {f1 [WITH {c | NULL}] | itab}.
```

Description

Option 1. (`f1`) Resets the contents of variable `f1` to the initial value of its type.

Option 2. (**`WITH`** `c`) Fills `f1` to its declared length with the first byte of `c`.

Option 3. (**`WITH NULL`**) Fills `f1` with hex 00. (Be careful: `NULL` is not a legal value for most Types – see `TYPE`).

Option 4A (`itab` with a header line) Resets all the fields of the header line to their initial values. The table is unchanged.

Option 4B. (`itab` with no header line) Empties the table.

See also `REFRESH`, `FREE`

CLOSE CURSOR

Syntax

```
CLOSE CURSOR cname.
```

Description

Explicitly closes the database cursor (record pointer) `cname`.

See also `OPEN CURSOR`, `FETCH`

CLOSE DATASET

Syntax

CLOSE DATASET filename.

Description

Explicitly closes the sequential file on the application server. Open files are implicitly closed at every screen change.

See also OPEN DATASET, READ DATASET, TRANSFER, DELETE DATASET

CNT

Syntax

CNT(keyfield).

Description

A system function available in LOOP structures on sorted dataset extracts where keyfield is in the sort key and is not a numeric field. Within the condition block AT LAST...ENDAT, **CNT** returns the number of distinct values of keyfield in the extract. Within the condition block AT END OF testfield ... ENDAT, **CNT** returns the number of distinct values of keyfield for the current value of testfield in the extract.

See also AT END OF, AT LAST, CNT, EXTRACT, FIELD-GROUPS, INSERT, LOOP, SUM()

CO

String comparison operator "Contains Only characters from".

See *Operators*

COLLECT

Syntax

COLLECT [wa **INTO**] itab.

Description

Compares all header line (or work area wa) fields other than types P, I & F to their corresponding fields in itab. If the combination of values is not found, **COLLECT** appends the header line into itab. If the combination was found, **COLLECT** sums header line P, I & F fields into the corresponding itab fields of the itab record in which the combination was found. SY-TABIX contains the record number of the new or modified record.

See also APPEND, INSERT, MODIFY, WRITE...TO

Color

See FORMAT, PRINT-CONTROL

Comments

Description

Comments are initiated by an asterisk (*) in the first column or by a double-quote (") in any column except in strings. Comments are terminated by the line break.

COMMIT

Syntax

```
COMMIT WORK [AND WAIT].
```

Description

Releases locks and table cursors, and executes a database commit. **COMMIT WORK** calls subroutines defined by the command `PERFORM...ON COMMIT`. It completes any update requests specified by `CALL FUNCTION...IN UPDATE TASK` and it executes work started by `CALL FUNCTION...IN BACKGROUND TASK`. Committed work cannot be reversed by `ROLLBACK WORK`. **AND WAIT** halts the program until all the type V1 (U1) updates are complete. Type V2 (U2) updates will be executed in parallel with the re-started program. Since table cursors are cleared by **COMMIT WORK**, this command shouldn't be used in a `SELECT... ENDSELECT` loop or while using database cursors (that is, `OPEN CURSOR` and `FETCH`). Debug automatically issues a `COMMIT WORK`, so the system state alters when you enter the debug mode.

SY-SUBRC	
	0 if successful
	>0 otherwise

See also `CALL DIALOG`

COMMUNICATION

Description

Use these several forms in the sequence shown to establish CPI-C communications between programs. The use of Remote Function Calls (RFCs) is preferred over this technique.

Syntax

```
COMMUNICATION INIT DESTINATION d1 ID id1 [RETURNCODE
rc1].
```

Requests connection with external system `d1`, assigns an ID number to the connection and assigns the return code to `rc1`.

```
COMMUNICATION ALLOCATE ID id1 [RETURNCODE rc1].
```

Allocates resources. This must immediately follow the **INIT** form.

COMMUNICATION ACCEPT ID idl **[RETURNCODE** rc1] **.**
External system d1 accepts the connection request.

COMMUNICATION SEND ID id1 **BUFFER** f1 **[RETURNCODE** rc1]
[LENGTH w1] **.**
Send the contents of f1 to the external system and specify the length of f1 to send in w1.

```
COMMUNICATION RECEIVE ID id1 BUFFER f1
 DATAINFO d1

 STATUSINFO s1
 INCLUDE 'RSCPICDF'
 [RETURNCODE rc1]
 [LENGTH w1]
 [RECEIVED w2]
 [HOLD] .
```

Receives the data from the external system into f1. **DATAINFO** returns information about the transmission. **STATUSINFO** returns information about the program status. **INCLUDE** 'RSCPICDF' so you can interpret the information provided by **DATAINFO** and **STATUSINFO**. **LENGTH** specifies the length of f1 to receive. **RECEIVED** populates w2 with the length of the data actually received. **HOLD** forces the program to wait for completion rather than rolling this out and performing other work; rolling out closes table cursors so SELECT loops would crash.

COMMUNICATION DEALLOCATE ID id1 **[RETURNCODE** rc1] **.**
Closes the connection and releases all resources.

COMPUTE

The keyword COMPUTE invokes arithmetic statements. Its use is unnecessary.

Example

```
COMPUTE length = STRLEN( arg ).
```

is the same as

```
length = STRLEN( arg ).
```

See *Operators*

CONCATENATE

Syntax

CONCATENATE a b c...**INTO** d **[SEPARATED BY** {e | **SPACE**}**].**

Description

Concatenates any number of fields. Operands are treated as type C without conversion. Trailing blanks are trimmed off before concatenation.

SEPARATED BY e inserts the string e between each of the operands.

SEPARATED BY SPACE inserts a space between each of the operands.

Example

```
DATA: st1(15) VALUE 'Texas',
      st2(15) VALUE 'California',
      st_list(32).
CONCATENATE st1 st2 INTO st_list SEPARATED BY ', '.
      st_list contains 'Texas, California'
```

SY-SUBRC = 0 if successful
= 4 if the result was too long for d; in that case it was copied to the length of d

See also the other string processing commands: CONDENSE, OVERLAY, REPLACE, SEARCH, SHIFT, SPLIT, STRLEN(), TRANSLATE

CONDENSE

Syntax

CONDENSE string **[NO-GAPS].**

Description

Moves all the "words" in string to the left until each is separated by one space, or **[NO-GAPS]** with no separation.

Example

```
DATA name(20) VALUE 'ABAP Computer Book'.
CONDENSE name. → name contains 'ABAP Computer Book'
CONDENSE name NO GAPS. → name contains 'ABAPComputerBook'
```

See also the other string processing commands: CONCATENATE, OVERLAY, REPLACE, SEARCH, SHIFT, SPLIT, STRLEN(), TRANSLATE

Condition

Description

Binary (that is, TRUE/FALSE) conditions are used in the tests CHECK, ELSEIF, IF, WHERE, WHILE and may take any of the following forms.

Condition	Description
f1 **[NOT]** OP v1	[Not] True if f1 and v1 satisfy the relation of OP. OP is any of the relational operators EQ, NE, GT, GE, LT, LE (see Operators, Relational). For the tests CHECK, ELSEIF, IF, WHILE (not including WHERE) OP may also be any of the string comparison operators CA, CO, CS, CP (see Operators, String). v1 may be SPACE.
[NOT] f1 **IS INITIAL**	[Not] True if the value of f1 is its initial value, which depends on its type. May be used in CHECK, ELSEIF, IF, WHILE.
f1 **[NOT] BETWEEN** v1 **AND** v2	[Not] True if the value of f1 is between the values of v1 and v2. Note: **AND** here does not refer to the logical product.
f1 **[NOT] LIKE** v1 **[ESCAPE** s1**]**	[Not] True if f1 contains a string that matches pattern v1. Construct a search pattern in which "_" represents one character and "%" represents any number of characters. A pattern character immediately following the escape character is interpreted literally. There is no default escape character.
f1 **[NOT] IN** (v1,..., vn)	[Not] True if f1 is found in any of the listed values v1- vn. Notice there is no space after opening parenthesis.
f1 **[NOT] IN** rtab	[Not] True if f1 satisfies any of the condition records in rtab, where that is an internal table like one created by RANGES. True for an empty rtab.
f1 **IS [NOT] NULL**	[Not] True if f1 contains **NULL** = hex 00. Null is <u>not</u> the initial value for most Types.
c1 **AND** c2	Logical product of two of the above conditions, true if both of c1 and c2 are true.
c1 **OR** c2	Logical sum of two of the above conditions, true if either of c1 or c2 are true.

Condition	Description
(itab)	Used only for WHERE statements in SELECT commands. itab contains condition statements in its single Type C field no wider than 72 characters. The statements must be literal, containing no variables. All the above conditions may be included except f1 **IN** rtab. For example: DATA: itab(72) OCCURS 10 WITH HEADER LINE. itab = 'COMPANY EQ ''2001''. APPEND itab. itab = 'COMPANY EQ ''2002''. APPEND itab.
c1 **AND** (itab)	You may specify both hard-coded conditions and itab conditions, again for SELECT only.
Compare two fields, bit-by-bit. b must be type X.	
a **O** b	(One) - True if all the '1's in b are also '1's in a.
a **Z** b	(Zero) - True if all the '1's in b are '0's in a.
a **M** b	(Mixed) - True if, of the '1's in b, at least one bit in a is '0' and at least one is '1'.

See also *Operators,* RANGES

CONSTANTS

Description

Declares and assigns values to global and local constants.

Syntax

For a single constant:

```
CONSTANTS fieldname[(length)]
  [TYPE datatype]
  [LIKE otherfieldname]
  [DECIMALS n]
  VALUE {lit | const | IS INITIAL }.
```

length applies only to types C, N, P, X. datatype may be any of the standard types, or types you have defined in **TYPES** commands; it defaults to type C (text). **LIKE** only inherits attributes, not values. **DECIMALS**" only applies to P type; n=0..14 and defaults to 0. const may be a system field such as SY-DATUM.

To declare several constants, use the colon-and-commas construction:

```
CONSTANTS:  fieldname1[(length)] options VALUE val1,
            fieldname2 VALUE val2,
            ...
```

To declare complex constants (constant arrays):

```
CONSTANTS:  BEGIN OF arrayname,
            f1 TYPE t1 VALUE val1,
            f2 TYPE t2 VALUE val2,
            ...
          END OF arrayname.
```

arrayname-f1 returns the unpacked value for f1, and arrayname returns all fields in a single packed string.

See also DATA, LOCALX, STATICS, TABLES, TYPES

CONTEXTS

Syntax

CONTEXTS c1.

Description

Declares the named context. You can then create a context instance with DATA mycontext TYPE CONTEXT_c1.

Populate the context keys using a SUPPLY statement and extract its dependent values with a DEMAND statement. A context is a Data Dictionary object you can create and display in the Context Builder SE33.

See also DEMAND, SUPPLY

CONTINUE

Syntax

CONTINUE.

Description

Unconditionally jumps to bottom of the current repetitive processing block for next iteration. **CONTINUE** is effective in DO, LOOP, SELECT, WHILE.

See also CHECK, EXIT, LEAVE, REJECT, STOP.

CONTROLS

Syntax

```
CONTROLS ctrl1
TYPE {TABLEVIEW USING SCREEN scr1 | TABSTRIP}.
```

Description

Declares a runtime control (an interactive screen) that you have previously defined. This may either be a table control (displaying multiple lines on the screen) or a tabstrip control (showing several tabs). You must write a `PBO` and a `PAI` to process the tabstrip tab selections or the table control in a step-loop. See the online help for far more information about setting these up.

See also `REFRESH CONTROL`

CONVERT

Syntax

```
CONVERT
{DATE d1 INTO INVERTED DATE d2
|INVERTED DATE d1 INTO DATE d2}.
```

Description

Converts between special defined formats. **`INVERTED DATE`** is the nine's-complement of the **`DATE`** internal representation `YYYYMMDD`, and therefore inverted dates sort in the reverse order to dates. This is rarely useful, as the **`DESCENDING`** option is available for sorting.

See *Type Conversions*

CONVERT *(text)*

Syntax

```
CONVERT TEXT txt1 INTO SORTABLE CODE sc1.
```

Description

Converts text string into a value that can be sorted alphabetically. The "text environment" encodes text strings depending on the current user language among other things. Converting such strings **`INTO SORTABLE CODE`** results in values that will sort alphabetically regardless of language or circumstances. Typically this command is used to create sortable fields in itabs or datasets. `txt1` must be type C. `sc1` must be type X and at least 16 (or 24 depending on which documentation you read) times as long as `txt1`, or type XSTRING, which will automatically be the correct length.

See also `GET LOCALE LANGUAGE`, `SET LOCALE LANGUAGE`

CONVERT *(timestamp)*

Syntax

```
CONVERT TIME STAMP ts1 TIME ZONE z1
INTO DATE d1 TIME t1.
CONVERT DATE d1 TIME t1 INTO TIME STAMP ts1
TIME ZONE z1.
```

Description

Converts values between a timestamp and a date and time for a particular time zone. Timestamp `ts1` must be type P8 or P11 with 7 decimal places, and Time Zone must be type C6. You can use Data Elements `TIMESTAMP` (P8), `TIMESTAMPL` (P11), and `TIMEZONE` in your DATA statement declaring those variables. Time stamp values are always in UTC (formerly Greenwich Mean Time) and are in the form YYYYMMDDHHMMSS. Time zone values are stored in the transparent table `TTZZ`.

`SY-SUBRC` 0 successful
4 time zone was initial; conversion did not consider time zone
8 time zone `z1` not found in table TTZZ
12 time stamp, date or time invalid for type; not converted

See also `GET TIME STAMP`

Conversion

See *Type conversions*

COS

Syntax

```
COS( y ).
```

Description

Returns the cosine of floating point number `y`, for `y` in radians.

See also *Arithmetic functions*

COSH

Syntax

```
COSH( y ).
```

Description

Returns the hyperbolic cosine of floating point number `y`.

See also *Arithmetic functions*

Country

See `SET COUNTRY`

`CP`

String comparison operator "Contains the Pattern".
See *Operators*

CPI-C

"Common Programming Interface – Communications" The protocol for synchronous inter-program communication among SAP products, used by the Gateway Server.
See also `COMMUNICATION`

`CREATE OBJECT` *(OLE)*

Syntax

```
CREATE OBJECT obj1 class1 [LANGUAGE lng1].
```

Description

Registers `obj1` with SAP so subsequent OLE2 operations can be executed. **`LANGUAGE`** specifies the language for the method and properties; the default is English.

`SY-SUBRC` 0 successful
1 communication error, described in `SY-MSGLI`
2 function call error
3 OLE API error
4 `obj1` not registered with SAP

See also `CALL METHOD`, `FREE OBJECT`, `GET PROPERTY`, `SET PROPERTY`

`CREATE OBJECT` *(Objects)*

Description

Creates an object in ABAP Objects.
This book doesn't discuss ABAP Objects programming.
See also *Object-oriented programming*

`CS`

String comparison operator "Contains the String".
See *Operators*

`CURSOR`

See `GET CURSOR`, `SET CURSOR` for list processing
See `CLOSE CURSOR`, `FETCH`, `OPEN CURSOR` for database processing

DATA

Description

Declares variables and optionally assigns their starting values and attributes.

Field syntax

To declare a single field:

```
DATA fieldname[(length)]
    [TYPE datatype]
    [LIKE otherfieldname]
    [DECIMALS n]
    [VALUE lit | const].
```

Where `length` applies only to types `C`, `N`, `P`, `X`. `datatype` may be any of the standard types, or types you have defined in **TYPES** commands; it defaults to type `C` (text). **LIKE** only inherits attributes, not values. **DECIMALS**" only applies to P type; `n=0..14` and defaults to `0`. **VALUE** defaults to the initial value. `const` may be a system field such as SY-DATUM.

To declare several fields, use the colon-and-commas construction:

```
DATA: fieldname1[(length)] options,
      fieldname2...
```

Array syntax

Use any of the following techniques to declare an array. `array1-f1` returns the unpacked value for `f1`. `array1` returns all fields in a single packed string.

```
DATA: BEGIN OF array1,
      f1 TYPE t1, f2 TYPE t2...
      END OF array1.
```

or where `array2` was previous defined in **DATA** or **TYPES**

```
DATA: array1 LIKE array2.
```

or where `itabtype` is a table type created by **TYPES**

```
DATA: array1 TYPE LINE OF itabtype.
```

or where `itab` is a table previously created by **DATA**

```
DATA: array1 LIKE LINE OF itab.
```

Itab syntax

Use the following technique to declare an internal table, specifying the kind of table.

```
DATA itab
{TYPE kind1 OF arraytype1 | LIKE kind2 OF array1 }
WITH {UNIQUE | [NON-UNIQUE]}
      {KEY k1 k2 kn | KEY TABLE_LINE | DEFAULT KEY}
[INITIAL SIZE n1] [WITH HEADER LINE].
```

Where `kind` is one of `[STANDARD TABLE]`, `SORTED TABLE`, `HASHED TABLE`. The standard table is the default and must be non-unique. The hashed table must be unique, and the sorted table may be either. Fields `k1`, `k2`... are the selected keys, searched in the order listed. **`TABLE_LINE`** is the entire line of the table as a string. **`DEFAULT KEY`** is the "standard" key definition which consists of all the table's fields which are not numeric (F, I, P) and not internal tables themselves, in the order in which they appear in the line.

Use any of the following techniques to declare an internal table with a header line.

```
DATA: itab LIKE array OCCURS nocc WITH HEADER LINE.
```

or where `itabtype` is a table type created by **`TYPES`**

```
DATA: itab TYPE itabtype OCCURS nocc WITH HEADER LINE.
```

or

```
DATA BEGIN OF itab OCCURS nocc.
   INCLUDE STRUCTURE s1. "s1 may be dbtab, itab or wa
DATA END OF itab.
```

(Note: the **`LIKE`** construction is preferred over **`INCLUDE STRUCTURE`**)

or

```
DATA BEGIN OF itab OCCURS nocc.
   INCLUDE STRUCTURE s1.    "s1 may be dbtab, itab or wa
   INCLUDE STRUCTURE s2.    "s2 may be dbtab, itab or wa
DATA END OF itab.
```

or

```
DATA BEGIN OF itab OCCURS nocc.
   INCLUDE STRUCTURE s1. "s1 may be dbtab, itab or wa
```

```
DATA: fieldname1[(length)] options,
      fieldname2.
DATA END OF itab.
```

or

```
DATA: BEGIN OF itab OCCURS nocc,
         f1 TYPE t1,
         ...
      END OF itab.
```

or

```
DATA: BEGIN OF itab OCCURS nocc,
         istruct LIKE dbtab,
      END OF itab.
```

(Refer to the fields using the construction: `itab-istruct-fieldname.`)

Use any of the following techniques to declare an internal table without a header line.

```
DATA: itab LIKE array OCCURS nocc.
```

or where `type` is any standard or complex type.

```
DATA: itab TYPE type OCCURS nocc.
```

or where `itabtype` is a table type created by **TYPES**

```
DATA: itab TYPE itabtype.
```

Use either of the following techniques to declare a RANGES table (see RANGES).

```
DATA: rtab LIKE RANGE OF f1 [INITIAL SIZE n1]
[WITH HEADER LINE].
```

or where `ty1` is any standard type.

```
DATA: rtab TYPE RANGE OF ty1 [INITIAL SIZE n1]
[WITH HEADER LINE].
```

The **OCCURS** value `nocc` must be a literal number; zero is allowed. Its value is effective only in the command `APPEND SORTED BY`. The later **INITIAL SIZE** parameter specifies the number of lines in the initial itab, and zero is permitted. In both cases, the itab size is dynamically adjusted in operations.

See also CONSTANTS, LOCAL, STATICS, TABLES, TYPES

Database Server

Description

The "bottom layer" in the SAP three-layer client-server-server architecture, where the SQL database resides. It is below the SAPGUI Presentation Server and the Application Server.

See also *Application Server, Presentation Server, SAPGUI*

Date

Description

Date type variables are stored as packed fields interpreted as the number of days since 01/01/0001. Addition and subtraction is supported between dates and type P and I numbers.

See also `CONVERT`, *Appendix A* for the system field `SY-DATUM`

dbtab

Description

A table maintained by SAP database that may contain system information, master data, lookup information or transaction data. The commands that affect `dbtabs` include:

`DELETE`, `DESCRIBE DISTANCE`, `FREE`, `GET`, `INSERT`, `MODIFY`, `SELECT`, `TABLES`, `UPDATE`.

See also *itab, Table Types*

Debug

Description

Start the debug mode with `/H` in the command field (that is, "Hobbel" mode). Debug automatically issues a `COMMIT WORK`, so the program state alters when you start it.

See also `BREAK-POINT` *and transactions* /`ST05` *(SQL Trace)*, /`SE30` *(ABAP Trace)*, /`SDBE` *(Explain SQL)*

`DEFINE`

Syntax

```
DEFINE macroname.
    ...                          "ABAP statements
END-OF-DEFINITION.
```

Description

This command pair defines a macro. The macro's code may include up to nine parameters `&1..&9` which you populate using calling arguments. Call the macro by using its name in your code, followed by any required arguments. Macros may be nested, but they may not recurs.

Example

```
DATA: title(32) VALUE 'Report Header', colhdr(72).
colhdr = 'Index      Name      Description    Date'.
DEFINE writeheader.
   ULINE.
   WRITE:   SY-DATUM, &1 CENTERED,
            SY-UNAME RIGHT-JUSTIFIED, /, &2.
END-OF-DEFINITION.
...
WRITE 'Before macro'.
writeheader title colhdr.
WRITE: / 'After macro'.
```

Results in:

```
Before macro
--------------------------------------------------------
05/06/1997              Report Header            USER01
Index    Name     Description    Date
After macro
```

DELETE ADJACENT DUPLICATES

Syntax

DELETE ADJACENT DUPLICATES FROM itab
[COMPARING { f1 f2...| **ALL FIELDS**} **].**

Description

If there are consecutive records in itab whose primary keys are equal, this command will delete the second of those records, and will continue comparing and deleting until the primary keys no longer match. The primary keys of an itab are all of the non-F I P fields. **COMPARING** deletes the second of adjacent records if their listed field values are equal. **COMPARING ALL FIELDS** deletes the second of adjacent records if their entire records are equal. This command is primarily useful after a SORT [DESCENDING] on itab.

SY-SUBRC 0 if at least one record is deleted
4 otherwise

DELETE ITAB

Syntax

DELETE itab
[INDEX ndx
| **WHERE** <condition>
| **[FROM** ndx1] **[TO** ndx2]]**.**

Description

Deletes the current record from itab in a LOOP..ENDLOOP structure. **INDEX** deletes record number ndx from itab. **WHERE** deletes records from itab that satisfy the condition. **FROM** starts deleting from record number ndx1 where ndx1 > 0. **TO** deletes down to and including record number ndx2 where ndx2 ≥ ndx1.

SY-SUBRC 0 if at least one record is deleted
4 otherwise

See also INSERT, MODIFY

DELETE DBTAB

Syntax

```
DELETE dbtab [CLIENT SPECIFIED]
[FROM {wa | TABLE itab}].
```

Description

Deletes one record from dbtab whose primary key matches that in the header line. The primary key consists of all the key fields. The command normally affects only the current client records. If you use **CLIENT SPECIFIED** then client (MANDT) becomes a normal field and you can affect other client's records. **FROM** wa deletes one record from dbtab whose primary key matches the left-most characters in the work area wa out to the length of the primary key. The structure of wa is ignored. **FROM TABLE** itab deletes all records from dbtab whose primary keys match those of the left-most characters in any record of itab out to the length of the primary key. The structure of itab is ignored.

SY-DBCNT contains the number of records deleted

SY-SUBRC 0 if at least one record was deleted or if itab was empty
4 if no primary key was matched

DELETE FROM DBTAB

Syntax

```
DELETE [CLIENT SPECIFIED] FROM dbtab WHERE <condition>.
```

Description

Deletes those records from dbtab that satisfy the **WHERE** condition. The command normally affects only the current client records. If you use **CLIENT SPECIFIED** then client (MANDT) becomes a normal field and you can affect other client's records.

SY-DBCNT contains the number of records deleted

SY-SUBRC 0 if at least one record was deleted or if itab was empty
4 if no primary key was matched

See also WHERE

DELETE DATASET

Syntax

DELETE DATASET filename.

Description

Deletes the named sequential file on the application server.

SY-SUBRC 0 if successful
>0 otherwise

See also CLOSE DATASET, OPEN DATASET, READ DATASET, TRANSFER

DELETE REPORT

Syntax

DELETE REPORT rpt1.

Description

Deletes the source code, attributes, textpool and generated version of program rpt1. The documentation and variants are not deleted. The function module RS_DELETE_PROGRAM is the preferred way to delete a report.

SY-SUBRC 0 if the program was deleted
>0 otherwise

See also DELETE TEXTPOOL, INSERT REPORT, READ REPORT

DELETE TEXTPOOL

Syntax

DELETE TEXTPOOL rpt1 **LANGUAGE** {lng1 | *}.

Description

Deletes the text elements for program rpt1 in language lng1, or (*) in all languages. SY-LANGU contains the language selected at login.

See also INSERT TEXTPOOL, READ TEXTPOOL, *Text Elements*

DEMAND

Syntax

DEMAND d1 = f1 d2 = f2 ... **FROM CONTEXT** c1.

Description

Assigns to the variables f1 f2... the dependent values in the named fields that correspond to the key fields of the context previously populated with a SUPPLY statement.

See also CONTEXTS, SUPPLY

DEQUEUE

See ENQUEUE

DESCRIBE DISTANCE

Syntax

DESCRIBE DISTANCE BETWEEN tab-f1 **AND** tab-f2 **INTO** nchar.

Description

Returns in nchar the number of characters between the beginning of field f1 and the beginning of field f2 in table tab where tab may be a dbtab, an itab, or an array.

DESCRIBE FIELD

Syntax

```
DESCRIBE FIELD f1
  [LENGTH len]
  [OUTPUT-LENGTH len]
  [DECIMALS n]
  [EDIT MASK mask]
  [TYPE typ [COMPONENTS n]].
```

Description

Returns the requested attributes about field f1. You must specify at least one attribute. **LENGTH** returns the length in len. **OUTPUT-LENGTH** returns the output-length in len. **OUTPUT-LENGTH** is the field length as it would be printed in a list, see its Domain definition. For example Type P's output length may be twice its length. **DECIMALS** returns the decimals in the DATA declaration of this type P field. **EDIT MASK** returns the name of f1's Dictionary conversion routine with a prefix of '=='. **TYPE** returns the type in typ. If f1 is a structure or itab then **COMPONENTS** returns in n the number of its components.

Values for **TYPE** may be any of the standard data types (that is, C, I, F, etc) or these special cases:

Type	Description
h	Internal table
s	2-byte integer with leading sign
b	1-byte integer without leading sign
u	Structure without internal table
v	Structure containing at least one internal table

DESCRIBE LIST

Syntax

```
DESCRIBE LIST
 [NUMBER OF {LINES ln1 | PAGES p1 }
 | LINE ln2 PAGE p2
 | PAGE p3
  [INDEX ndx1]
```

```
    [LINE-SIZE w1]
    [LINE-COUNT r1]
    [LINES n1]
    [FIRST-LINE n2]
    [TOP-LINES n3]
    [TITLE-LINES n4]
    [HEAD-LINES n5]
    [END-LINES n6] ]
  [INDEX ndx2].
```

Returns the following attribute(s) of the current list. **NUMBER OF LINES** returns in `ln1` the number of lines. **NUMBER OF PAGES** returns the number of pages in the current list in `p1`. **LINE** `ln2` **PAGE** `p2` returns in `p2` the page number of line `ln2`. **INDEX** `ndx2` returns the current list level. **PAGE** returns the following attributes of page `p3`:

INDEX ndx1	List level of the page
LINE-SIZE	Line width for the page
LINE-COUNT	Maximum number of lines allowed
LINES	Number of lines written on the page
FIRST-LINE	Line number of first line
TOP-LINES	Number of lines in page header (title, column headers & TOP-OF-PAGE)
TITLE-LINES	Number of title lines
HEAD-LINES	Number of column header lines
END-LINES	Number of lines reserved for end of page processing

SY-SUBRC 0 if successful
>0 index level doesn't exist (**NUMBER OF … INDEX** only)
4 line doesn't exist (**LINE…PAGE** only)
4 page doesn't exist (**PAGE** only)
8 list doesn't exist (**LINE…PAGE** & **PAGE** only)

DESCRIBE TABLE

Syntax

DESCRIBE TABLE `itab` **[LINES** `nlin`**]** **[OCCURS** `nocc`**]**.

Description

LINES returns in `nlin` the current number of records in `itab`. The system field SY-TFILL also contains the number after this command is executed. **OCCURS** returns in `nocc` the occurs value of `itab`. The system field SY-TOCCU also contains the number after this command is executed. Note: SAP may change the occurs value from the DATA value during execution. You must specify at least one attribute.

See also OCCURS

Dispatcher

The dispatcher is a central controller in every SAP product that continuously routes work sessions among the appropriate Work processes established in the instance.

See *Instance, Work processes*

`DIV`

Integer division operator.

See *Operators, Arithmetic*

`DIVIDE`

Syntax

```
DIVIDE a BY b.
```

Description

Equivalent to: `a = a / b.` Division by `0` is illegal, except `0 / 0 = 0.`

Non-numeric fields are converted. See "Type Conversions" below for conversion information.

See also `ADD`, `MULTIPLY`, `SUBTRACT`

`DIVIDE-CORRESPONDING`

Syntax

```
DIVIDE-CORRESPONDING array1 BY array2.
```

Description

If `array1` & `array2` are structured work areas such as header lines, then this command divides like-named fields in `array1` and `array2`.

Example

```
DIVIDE array1-key1 BY array2-key1.
DIVIDE array1-key2 BY array2-key2.
etc.
```

See also `ADD-CORRESPONDING`, `MULTIPLY-CORRESPONDING`, `SUBTRACT-CORRESPONDING`

`DO`

Syntax

```
DO [n TIMES] [VARYING v1 FROM array1-fm NEXT array1-fn].
  ..
ENDDO.
```

Description

Loops `n` times or until terminated by **`EXIT`** or **`STOP`**. Changing `n` inside the loop doesn't affect the number of times the loop will execute. **`VARYING`** steps the variable v1 in subsequent passes, see `VARY` for an example.

CONTINUE unconditionally jumps back to the **DO** for the next iteration. **CHECK** skips to **ENDDO** for the next iteration if the condition is false. **EXIT** terminates the loop; execution continues with the statement after **ENDDO**. **STOP** terminates the loop; execution continues with the event **END-OF-SELECTION**. **DO**-loops and WHILE-loops may be nested indefinitely. SY-INDEX contains the one-based current step for the current nest level; after the **ENDDO** SY-INDEX is restored to its value before the **DO**.

See also CASE, IF, LOOP, VARY, WHILE

DOWNLOAD

Description

Function Module to write an itab to a local disk file on the user's workstation. This function module presents the user with a dialog box for the user to enter filename and type (ASCII, BIN, Excel/DAT, spreadsheet/WK1).

Hint: to insert function module calls into your program, use [Pattern. See WS_DOWNLOAD for details; it is similar except filename and type are parameters rather than prompts.

See also UPLOAD, WS_DOWNLOAD, WS_EXECUTE, WS_QUERY, WS_UPLOAD

Duplicate records

See DELETE ADJACENT DUPLICATES

Dynpros

"Dynamic Programs" refer to user transactions, and include their graphical screen definitions and their associated flow control code.

EDI

"Electronic Data Interface", communications protocol between an SAP system and an external system. EDI typically uses ANSI X112 or EDIFACT data formats over a Value-Added Network (VAN). The SAP system uses IDOCS as the data vehicle to a third-party translator.

See also *ALE, IDOCS*

EDITOR-CALL

Syntax

EDITOR-CALL FOR {itab [**TITLE** 'text string'] | **REPORT** rpt1} [**DISPLAY MODE**] .

Description

Places the internal table itab or the program rpt1 in the SAP editor so the user may edit it. In **DISPLAY MODE** the user may only view it. itab must contain only type C fields; the record length is limited to 72 characters. The *Save* button `F11` saves changes and returns to the calling program. The *Return* button `F3` closes the editor without saving.

`SY-SUBRC` 0 if changes were saved before leaving
4 otherwise

END-OF-PAGE

Syntax

`END-OF-PAGE.`

Description

Event triggered at the end of each basic list and detailed list page if an end-of-page area is reserved by the `LINE-COUNT` parameter (in the `REPORT` statement or in a previous `NEW-PAGE` statement) or by a `RESERVE` statement. **`END-OF-PAGE`** is not triggered at the end of the list unless the line count is full. `NEW-PAGE` does not trigger this event.

See also *Events*, `TOP-OF-PAGE`

END-OF-SELECTION

Syntax

`END-OF-SELECTION.`

Description

Event triggered after all the Logical Database records have been processed, when the `START-OF-SELECTION` event completes if no Logical Database is in use, and by the `STOP` command.

See also *Events*

ENQUEUE_OBJECTNAME...

Description

A set of function modules used to lock objects. Locks are automatically released at the end of the transaction; they can also be released by calling the matching `DEQUEUE_OBJECTNAME...` function module.

EQ

Relational operator "Equal".

See *Operators*

EVENTS

Description

Defines events of an ABAP Objects class or interface.

This book doesn't discuss ABAP Objects programming.

See also *Object-oriented programming*

Events

Description

When any event is triggered, processing jumps to the first command following the event statement, and continues to the end of that code block. The code block is terminated by the next event statement, the first `FORM` definition, or the end of the program.

See also `CHECK`, `STOP`

Events in on-line transactions:

```
AT LINE-SELECTION.
AT PFnn.
AT USER-COMMAND.
PROCESS AFTER OUTPUT.
PROCESS BEFORE INPUT.
PROCESS ON HELP-REQUEST.
PROCESS ON VALUE-REQUEST.
```

Events in reports & programs:

```
AT SELECTION-SCREEN [ON p | ON s | OUTPUT].
END-OF-PAGE.
END-OF-SELECTION.
GET [LATE].
INITIALIZATION.
START-OF-SELECTION.
TOP-OF-PAGE [DURING LINE SELECTION].
```

The normal order of events in a program is:

`INITIALIZATION` one time before the selection screen is displayed,

`AT SELECTION-SCREEN OUTPUT` every time before the selection screen is shown,

`AT SELECTION-SCREEN ON` `{p | s}` if user has specified the parameter `p` or select-option `s` in a screen,

`AT SELECTION-SCREEN` when the user accepts the screen, then the implicit start-of-selection code between the **`REPORT`** statement and the first event statement,

`START-OF-SELECTION` code block, and if `REPORT DEFINING DATABASE ldb` the internal LDB reader steps through the logical database hierarchically, one record at a time, triggering the appropriate **`GET`** events as the records are available,

and finally

`END-OF-SELECTION`.

EXEC SQL

Syntax

```
EXEC SQL [PERFORMING formname].
   ...
ENDEXEC.
```

Description

Provides the ability to issue native SQL commands directly to the database. This is dangerous business. Do it only if (1) it's absolutely necessary, and (2) you really <u>really</u> know what you're doing! **PERFORMING** calls the subroutine `formname` after each record retrieved by the SQL statement.

See also `EXIT FROM SQL`, *Open SQL*

Execute a program

To execute an ABAP from your terminal use /SA38, /SE 38 or /SE37. To call another ABAP program or report from within an ABAP see `SUBMIT`. To launch an operating system (that is, UNIX or Windows NT) program on the application server from within an ABAP see `OPEN DATASET...FILTER`, and `/SM49`. To launch a program on the presentation server (local workstation) see `WS_EXECUTE`.

EXIT

Syntax

```
EXIT [FROM { STEP-LOOP | SQL } ].
```

Description

Terminates the processing blocks **AT..., AT...ENDAT, DO, END-OF-PAGE, FORM, FUNCTION, LOOP, MODULE, [, TOP-OF-PAGE, WHILE**; outside of those blocks, cancels report processing and displays whatever list had been generated to that point;

See also **CHECK, CONTINUE, LEAVE, REJECT, STOP**

FROM STEP-LOOP - in Flow Control, departs current screen, ceases processing of the PBO and skips the PAI

See also *Flow Control*

FROM SQL - departs the loop created by **EXEC SQL PERFORMING** `formname`

EXP

Syntax

EXP(y).

Description

Returns the exponential of floating point number y; that is, **e****y for **e** =2.7182818284590452...

See also *Arithmetic functions*

EXPORT

Syntax

```
EXPORT f1 [FROM g1] f2 [FROM g2]...
TO {MEMORY | {DATABASE | SHARED BUFFER} tab(a)
[CLIENT cl1]}
ID ident.
```

Description

Stores the name and value of the listed data objects. The objects can be simple fields, records, table work areas and internal tables (but not their header lines). **TO MEMORY** stores them in user-defined memory files. **TO DATABASE** stores them in the named database table in cluster a. **TO SHARED BUFFER** stores them in the named cross-application buffer in cluster a. You must have declared the table in a TABLES statement, and it must have a standardized structure. Have a look at the dictionary table INDX for an example of the structure. f1 **FROM** g1 stores the data object g1 under the name f1. You can supercede the normal client processing by specifying a different client. **ID** associates the key or label ident with this group of objects. Each **EXPORT** to an ident over-writes the previous such **EXPORT**. The data objects can be retrieved (using the names f1..) with **IMPORT** and the same ident. The user-assigned memory is released when the current call chain is completed.

See also, FREE MEMORY, IMPORT FROM DATABASE, IMPORT FROM MEMORY, IMPORT FROM SHARED BUFFER

EXTRACT

Syntax

EXTRACT fg.

Description

Temporarily stores in memory as a sequential record the field group fg and its header. Records defined by different field groups can be interleaved. These interleaved records can be sorted and then analyzed in a LOOP ... ENDLOOP code block.

See also [, FIELD-GROUPS, INSERT, LOOP, SORT, SUM()

False

Description

There's no logical type in ABAP. Logical state is frequently represented by a 1-character TYPE C field with its initial value ' ' or `SPACE` for `FALSE`, and `'X'` for `TRUE`.

FETCH

Syntax

```
FETCH NEXT CURSOR cname INTO wa.
```

Description

Fills `wa` with the contents of the record identified by `cname`, then increments `cname`. The cursor `cname` must be of `TYPE CURSOR` and previously created with an `OPEN CURSOR` command. Table cursors are cleared by `COMMIT WORK`, so any commands that invoke it shouldn't be used before a **`FETCH`** command. Those commands include `CALL SCREEN`, `CALL TRANSACTION`, `COMMIT WORK`, `BREAKPOINT`.

`SY-SUBRC`	0 if a line was read 4 otherwise
`SY-DBCNT`	number of lines read using cursor `cname`

See also `CLOSE CURSOR`, `OPEN CURSOR`, `SELECT`

Field names

See *Names*

FIELD-GROUPS

Syntax

```
FIELD-GROUPS fg.
```

Description

Declares the field group `fg`, which is a set of fields selected from one or more tables. Assign the fields to `fg` using `INSERT`, fill field group `fg` from the tables using `EXTRACT`, and analyze the extracted dataset with `LOOP ... ENDLOOP`.

See also `AT...ENDAT`, `CNT`, `EXTRACT`, `INSERT`, `LOOP`, `SORT`, `SUM()`

FIELD-SYMBOLS

Syntax

```
FIELD-SYMBOLS <fs> [<type> | STRUCTURE dbtab DEFAULT wa].
```

Description

Declares the symbolic field `<fs>`. The program may then assign an actual field or pointer at runtime using the `ASSIGN` command. `<type>` is a literal type declaration. **`STRUCTURE`** assigns the structure of `dbtab` to `fs`. Refer

to the elements as `fs-f1` and so forth. **DEFAULT** sets wa as the initial assignment target of `fs`.
See also *Appendix G.2*, ASSIGN

FLOOR

Syntax

```
FLOOR( x ).
```

Description

Returns the floor of any number x, that is, the largest integer not greater than x, the integer value of x.
See also *Arithmetic functions*

Flow Control

See CHAIN, *Events*, EXIT FROM STEP-LOOP, MODULE, PROCESS...

FORM

Syntax

```
FORM name
[TABLES itab1 [STRUCTURE dbtab]]
[USING {[VALUE] [wa STRUCTURE dbtab]
| [f1 [TYPE t1]]...}]
[CHANGING {[VALUE] f2 [LIKE v1]...}].
  ...
ENDFORM.
```

Description

Defines a subroutine. Parameters are optional; formal parameters must have matching actual parameters in the calling **PERFORM** statement. Formal parameters with **TYPE** or **LIKE** clauses must match their actual parameter types. **USING** and **CHANGING** pass variables by reference. **VALUE** passes variables by value. **TABLES, USING** and **CHANGING** must appear in that order. You normally will pass parameters to the subroutine with **USING** and return them with **CHANGING**. You must place source code for forms at the end of the program. Definitions may not nest, that is, one **FORM** may not be defined inside another **FORM** definition. Calls may nest and recurse. **EXIT** immediately terminates the subroutine normally. **CHECK** <condition> terminates the subroutine normally if the condition is false. **STOP** unconditionally jumps to the **END-OF-SELECTION** event. An error message terminates the subroutine abnormally.
See also PERFORM

FORMAT

Syntax

```
FORMAT
[ COLOR [n | OFF]
| HOTSPOT [ON | OFF | var]
| INTENSIFIED [ON | OFF | var]
| INVERSE [ON | OFF | var]
| INPUT [ON | OFF | var]
| RESET ].
```

Description

Sets or modifies the screen output. Printer output is affected by PRINT CONTROL. **ON** is the default, so **INPUT ON** is the same as **INPUT**. Use var = 0 for **OFF** and var > 0 for **ON**; var should be type I. **HOTSPOT** turns the mouse cursor to pointing hand in this field. **INTENSIFIED** changes the background from pastel to saturated color. **INVERSE** exchanges the foreground and background colors. **INPUT** permits entering data into this field. **RESET** returns all attributes to **OFF**.

Color Parameter	Color
OFF | col_background	GUI-specific (default)
1 | col_heading	grayish blue
2 | col_normal	bright grey
3 | col_total	yellow
4 | col_key	bluish green
5 | col_positive	green
6 | col_negative	red
7 | col_group	violet

FORMAT commands take affect with the next WRITE or NEW-LINE; each new event resets **COLOR**, **INTENSIFIED**, **INVERSE**, **INPUT**. Use READ LINE to accept the data entered in **INPUT** fields.

See also PRINT CONTROL, WRITE

Formfeed

See NEW-PAGE

FRAC

Syntax

FRAC(x).

Description

Returns the fractional part of any number x, that is, the portion of x to the right of the decimal point.

See also *Arithmetic functions,* FLOOR, TRUNC

FREE

Syntax

FREE {f | itab | dbtab | **MEMORY [ID** ident]}.

Description

Clears data object f to its initial value, and releases allocated resources.
Empties itab and frees the memory allocated; it remains defined and can be reused. If itab has a header line, the header line remains unchanged.
Releases the work area for dbtab; it remains defined and can be reused.
Releases memory allocated for EXPORT TO MEMORY.

See also CLEAR, REFRESH

FREE OBJECT

Syntax

FREE OBJECT obj1.

Description

Releases the memory allocated to the OLE2 object obj1.

SY-SUBRC 0 memory was successfully released
1 communication error, described in SY-MSGLI
2 function call error

See also CALL METHOD, CREATE OBJECT, GET PROPERTY, SET PROPERTY

FUNCTION

Syntax

FUNCTION f1.

Description

Begins definition of the function module f1.

See also *Function Group, Function Module*, FUNCTION-POOL

Function Group

Description

To create a function group follow the path:

/SE37 {Goto {Function groups {Create group <name – 'Z*' ...

SAP will create a program shell named SAPLZ* containing an INCLUDE named LZ*TOP headed by the FUNCTION-POOL statement. Use that INCLUDE to store global data for the function group.

Function Module

Description

To create a function module follow the path:

/SE37 <name = 'Z_*' [Create...

(name is limited to 30 characters which may include letters & the underscore). SAP will create an INCLUDE in SAPLZ* which is headed by FUNCTION name; the function module is invoked by CALL FUNCTION.

FUNCTION-POOL

Syntax

```
FUNCTION-POOL fp1.
```

Description

The header statement of source code of the function modules in the function group.

See also *Function Module*, PROGRAM, REPORT

Gateway server

An instance of the SAP system.

See also *Instance*

GE

Relational operator "Greater than or Equal".

See *Operators*

GET

Syntax

```
GET dbtab [LATE].
```

Description

Logical database event triggered by the internal logical database reader program: the next record of dbtab is available. The **LATE** event is triggered when the record in dbtab is going to change (that is all lower level tables have been processed).

See also *Appendix G.3*, *Events*, REJECT, ON CHANGE OF

GET BIT

Syntax

GET BIT n1 **OF** f1 **INTO** g1.

Description

Assigns to g1 the bit in field f1 in (one-based) bit position n1. f1 must be type X and n1 must be greater than zero.

SY-SUBRC 0 success

>0 n1 is greater than the length of f1

See also SET BIT

GET CURSOR FIELD

Syntax

GET CURSOR FIELD f1
[**LENGTH** len]
[**LINE** lin]
[**OFFSET** off]
[**VALUE** val].

Description

In list processing **GET CURSOR FIELD** returns in f1 the global field name where the cursor is currently located. It also returns these other requested values. len is the output length of the field. lin is the absolute line number in the list, also shown in SY-LILLI. off is the zero-based position of the cursor in the field. val is the screen contents of the value of the field (shown as characters). If the cursor is located on a screen literal, local variable or VALUE parameter of a subroutine this command will return f1 = SPACE and val = 0.

SY-SUBRC 0 if the cursor is positioned on a field

4 otherwise

See also SET CURSOR

GET CURSOR LINE

Syntax

GET CURSOR LINE lin
[**LENGTH** len]
[**OFFSET** off]
[**VALUE** val].

Description

In list processing **GET CURSOR LINE** returns in lin the absolute line number in the list where the cursor is currently located, the same value shown in SY-LILLI. It also returns these other requested values. len is the output length of the line, which is the value established in the LINE-SIZE option of the REPORT and NEW-PAGE commands. off is the zero-based position of the cursor in the list line. `val` contains the screen contents of the value of the line (shown as characters) and doesn't include the `HIDE` fields.

`SY-SUBRC` 0 if the cursor is positioned on a list line
4 otherwise

See also `SET CURSOR`

GET LOCALE LANGUAGE

Syntax

`GET LOCALE LANGUAGE lg COUNTRY c MODIFIER m.`

Description

Assigns to `lg`, `c` and `m` the values of the current text environment. The fields must be type C with lengths found in table TPC0C.

See also `CONVERT` (*text*), `SET LOCALE LANGUAGE`

GET PARAMETER ID

Syntax

`GET PARAMETER ID key FIELD f1.`

Description

Assigns to `f1` the value of PID `key` in the user's SPA/GPA memory area.

`SY-SUBRC` 0 if a value was found for PID `key` (release 4.0 and above)
4 otherwise

See also `SET PARAMETER ID`

GET PROPERTY

Syntax

`GET PROPERTY OF obj1 p1 = f1 [NO FLUSH].`

Description

Assigns to field `f1` attribute `p1` of the OLE2 object `obj1`. `NO FLUSH` continues OLE2 bundling even if the next statement isn't an OLE2 command.

`SY-SUBRC` 0 all OLE2 commands in the bundle were successful
1 communication error, described in `SY-MSGLI`
2 method call error, described in dialog box
3 property set up error, described in dialog box
4 property read error, described in dialog box

See also `CALL METHOD`, `CREATE OBJECT`, `FREE OBJECT`, `SET PROPERTY`

GET RUN TIME

Syntax

`GET RUN TIME FIELD f.`

Description

Assigns to field `f` the time in microseconds since the first time this command was issued. `f` must be Type I. Use /SE30 to analyze complex runtime processes.

GET TIME

Syntax

`GET TIME [FIELD f1].`

Description

Resets `SY-DATUM` and `SY-UZEIT` to current system date and time and refreshes the local time zone fields `SY-TIMLO`, `SY-DATLO`, `SY-TSTLO` and `SY-ZONLO`. **FIELD** `f1` assigns the current system time to `f1` without changing `SY-UZEIT` or the time zone fields.

GET TIME STAMP

Syntax

`GET TIME STAMP FIELD f1.`

Description

Assigns to `f1` the current UTC (formerly Greenwich Mean Time) date and time in timestamp format. `f1` must be type P8 or P11 DECIMALS 7. The value of the P8 field will be in the format YYYYMMDDHHMMSS. The P11 field will add up to 7 decimal places of resolution to the seconds.

See also `CONVERT` *(timestamp)*, `WRITE...TIME ZONE`

GOTO

There is no GOTO-type command and there are no line labels in ABAP.

GPA

See `GET PARAMETER`, *SPA/GPA Memory Area*

GT

Relational operator "Greater Than".

See *Operators*

Header line

A Header Line is a runtime array that matches the structure of a database or internal table. In most cases, the system automatically moves data between the table and the Header Line, and you do most of your table-related work in the Header Line.

When you declare a database table in a `TABLES` statement, SAP automatically defines a Header Line with the same name as the table. At that time you may also declare an additional independent work area by including in the `TABLES` statement the tablename preceded by an asterisk.

When you create an `itab` with `BEGIN OF...INCLUDE STRUCTURE` or `LIKE... WITH HEADER LINE`, the system automatically creates a Header Line with the same name as the `itab`.

You can manually define a work area in a `DATA` statement using the `INCLUDE STRUCTURE tablename` or `LIKE tablename` constructions. Or finally you can (oh my!) manually define a work area in a `DATA` statement by explicitly calling out all the fields with their widths and types.

Non-table commands such as `MOVE` refer to the table's header line; table commands such as `APPEND` refer to the table itself, and sometimes to the header line. The notation `tablename[]` always refers to the table itself, not to the Header Line.

See also *Work area*

HIDE

Syntax

`HIDE f1 [f2...].`

Description

Stores `f1 [f2...]` in a "hidden" (that is, off-screen) area linked to the list lines created by an immediately-preceding **`WRITE`** statement. The variable may be a structure such as a header line, but that structure may not include a table. When an on-screen line is selected, the values in that line are assigned to the variables from which they came, including the hidden fields and the header line. **`HIDE`** is generally used to store key fields but there is no restriction to what it can store. The hidden field(s) need not have been included in the preceding **`WRITE`** statement.

Highlight a string

See SEARCH … AND MARK

IDOC

"Intermediate Document". The SAP standard data format for communicating with external systems (using EDI or EDIFACT) or internal SAP systems (using ALE).
See *ALE, EDI*

IF

Syntax

```
IF <condition1>.
    ..
[ELSEIF <condition2>.
    ..]
[ELSE.
    ..]
ENDIF.
```

Description

Finds the first true condition and executes the statements between that clause and the next **ELSEIF**, **ELSE**, or **ENDIF**, whichever it encounters first. Any number of **ELSEIF** statements are allowed. **ELSE** is optional and must follow the last **ELSEIF** statement. Each clause must be terminated with a period. **IF** statements may be nested without limit.
See also CASE, *Condition*, DO, WHILE

IMG

"Implementation Guide". The area in SAP that contains the configuration tables, and in which you can maintain configuration logs.
{Tools {Business Engineering {Customizing

IMPORT DIRECTORY

Syntax

```
IMPORT DIRECTORY INTO itab1 FROM DATABASE db1(area1)
  [CLIENT cl1] ID ident [TO wa1].
```

Description

Retrieves from the named area in the data cluster the directory of data objects having the ID key. The objects can be simple fields, records, table work areas and internal tables. **CLIENT** retrieves the directory from the named client in the case of client-dependent clusters. **TO** assigns the values to the named work area that has the same structure as the data cluster.

SY-SUBRC 0 successful
>0 otherwise

IMPORT FROM DATABASE

Syntax

```
IMPORT {(itab1) |
(f1 [= g1 | TO g1] f2 [= g2 | TO g2]...)}
  FROM DATABASE db1(ar) [CLIENT cl1] ID ident
  [TO wa1].
```

Description

Retrieves from cluster `ar` in the named database table the values of data objects `f1, f2,...`having the specified ID key. The objects can be simple fields, records, table work areas and internal tables. `itab1` contains a list of data objects and target fields in two text columns, one itab record per data object. = and **TO** assigns the memory data object `f1` to the field `g1`. **CLIENT** retrieves the data from the named client in the case of client-dependent clusters. **TO** assigns the values to the named work area that has the same structure as the data cluster.

SY-SUBRC 0 successful
>0 otherwise

See also EXPORT

IMPORT FROM MEMORY

Syntax

```
IMPORT {(itab1) |
(f1 [= g1 | TO g1] f2 [= g2 | TO g2]...)}
  FROM MEMORY ID ident.
```

Description

Retrieves from user-assigned memory the values of data objects `f1, f2,...`having the specified ID key. The objects can be simple fields, records, table work areas and internal tables. `itab1` contains a list of data objects and target fields in two text columns, one itab record per data object. **=** and **TO** assigns the memory data object `f1` to the field `g1`.

SY-SUBRC 0 if successful
>0 otherwise

See also EXPORT

IMPORT FROM SHARED BUFFER

Syntax

```
IMPORT {(itab1) |
(f1 [= g1 | TO g1] f2 [= g2 | TO g2]...)}
  FROM SHARED BUFFER db1(ar) [CLIENT cl1] ID ident
  [TO wa1].
```

Description

Retrieves from cluster `ar` in the cross-application buffer the values of data objects `f1, f2,...`having the specified ID key. The objects can be simple fields, records, table work areas and internal tables. `itab1` contains a list of data objects and target fields in two text columns, one itab record per data object. **=** and **TO** assigns the memory data object `f1` to the field `g1`. **CLIENT** retrieves the data from the named client in the case of client-dependent clusters. **TO** assigns the values to the named work area that has the same structure as the data cluster.

SY-SUBRC 0 successful
>0 otherwise

See also EXPORT

INCLUDE

Syntax

```
INCLUDE progname
```

Description

Includes the external program `progname` in the current program for syntax tests and operation. The entire **`INCLUDE`** statement must appear alone on one line. The location of the statement in your program depends on its contents. If it contains DATA statements, it should be located where the programs DATA statements are normally located. If it contains FORMs, then it should be located where the programs FORMs should be (that is, at the end). You may nest **`INCLUDE`** programs, that is, one **`INCLUDE`** may call another. Program `RSINCL00` can create a list of **`INCLUDE`**d programs and expand their code.

INCLUDE STRUCTURE s1.

See `DATA`

INFOTYPES

Syntax

```
INFOTYPES nnnn.
```

Description

An infotype is a structure in the data dictionary (a table structure containing no data) which is used by the Human Resources module. It is not used in general ABAP programming. HR programming is a specialty area that requires its own techniques. This command establishes a link to the named data dictionary object, and declares an itab with the structure of that object.

Example

```
DATA BEGIN OF Pnnnn OCCURS noccl.
  INCLUDE STRUCTURE Pnnnn.
DATA END OF Pnnnn VALID BETWEEN date1 AND date2.
```

Infotypes are grouped into the following categories:

nnnn	Type of Structure
0000 - 0999	HR Master data types
1000 - 1999	HR Planning data types
2000 - 2999	HR Time tracking data types
3000 - 8999	(unused to date)
9000 - 9999	Customer-defined data types

Use `SHOW INFOTYPES` in the ABAP editor to see the list of types and use `SHOW INFOTYPES nnnn` to see details about that specific structure.

See also `PROVIDE`

Initial values

See IS INITIAL

INITIALIZATION

Syntax

INITIALIZATION.

Description

Event triggered one time before the selection screen of the program or its logical database is displayed. You may set the PARAMETERS and SELECT-OPTIONS defaults at this point.

See also *Events* for their triggering order

INSERT DBTAB

Syntax

INSERT {dbtab | *dbtab | (dbtabname)}
[CLIENT SPECIFIED]
[FROM wa] **.**

Description

Inserts a new record into dbtab. dbtab takes the data from the header line. *dbtab takes the data from the work area *dbtab. dbtabname is a variable containing the name of the table. The command normally affects only the current client records. If you use **CLIENT SPECIFIED** then client (MANDT) becomes a normal field and you can affect other client's records. **FROM** wa takes the data from work area wa following the structure of dbtab; wa must be at least as wide as dbtab. Duplicate keys and existing UNIQUE indexes are not inserted.

SY-SUBRC 0 if the record was inserted
4 if key violation

SY-DBCNT contains the number of records inserted (0 or 1)

See also DELETE, MODIFY, UPDATE

INSERT DBTAB FROM TABLE

Syntax

`INSERT dbtab [CLIENT SPECIFIED] FROM TABLE itab [ACCEPTING DUPLICATE KEYS].`

Description

Inserts all the records from internal table `itab` into `dbtab`, following the structure of `dbtab`; `itab` must be at least as wide as `dbtab`. The command normally affects only the current client records. If you use **CLIENT SPECIFIED** then client (`MANDT`) becomes a normal field and you can affect other client's records. Key violations produce runtime errors. **ACCEPTING DUPLICATE KEYS** skips key violations without producing a runtime error, inserts the remaining records in `itab`, and sets `SY-SUBRC` to 4.

`SY-SUBRC` 0 all requested records were inserted or `itab` was empty
4 key violation and **ACCEPTING DUPLICATE KEYS** is chosen

`SY-DBCNT` contains the number of records inserted

See also `DELETE`, `MODIFY`, `UPDATE`

INSERT INTO DBTAB

Syntax

`INSERT INTO dbtab [CLIENT SPECIFIED] VALUES wa.`

Description

Equivalent to **INSERT** `dbtab` **FROM** `wa.`

INSERT ... INTO fg.

Syntax

`INSERT f1 [f2...] INTO fg.`

Description

Defines field group `fg` by inserting the listed field(s) from one or more tables. You declare the field group using `FIELD-GROUPS`, fill the field group from the tables using `EXTRACT`, and analyze the extracted dataset with `LOOP ... ENDLOOP`.

See also `CNT`, `EXTRACT`, `FIELD-GROUPS`, `LOOP`, `SUM()`

INSERT ... INTO itab

Syntax

```
INSERT
[INITIAL LINE
| LINES OF itab2 [FROM ndx1] [TO ndx2]
| wa ]
INTO itab1
[INDEX ndx].
```

Description

Inserts a new record into `itab` from the header line, or - `wa` - from work area `wa` at the current table cursor, pushing down the current record. **`INITIAL LINE`** inserts a new record with all fields initialized. **`LINES OF`** `itab2` inserts all the records in `itab2`. **`FROM`** starts from record number `ndx1` of `itab2` where `ndx1 > 0`. **`TO`** inserts down to and including record number `ndx2` of `itab2` where `ndx2` ≥ `ndx1`. **`INDEX`** inserts the new record into record number `ndx`, pushing the former record `ndx` to position `ndx+1`. If `ndx` > the number of records in the table + 2 then the record is not inserted.

`SY-SUBRC` 0 **`INDEX`** option was not used and the record was inserted
4 the index was too large

See also `DELETE`, `MODIFY`, `UPDATE`

INSERT REPORT

Syntax

```
INSERT REPORT rpt1 FROM itab.
```

Description

Inserts into the library the source code of program `rpt1`. The program name `rpt1` may be up to 30 characters long. `itab` must be no wider than 72 characters.

`SY-SUBRC` 0 the program was inserted
non-zero otherwise

See also `INSERT TEXTPOOL`, `DELETE REPORT`, `READ REPORT`

INSERT TEXTPOOL

Syntax

`INSERT TEXTPOOL rpt1 FROM itab LANGUAGE lng1.`

Description

Assigns the contents of `itab` to the text elements for program `rpt1` in language `lng1` and adds them to the library; see **READ TEXTPOOL** for the structure of `itab`.

`SY-SUBRC` 0 the textpool was inserted
non-zero otherwise

`SY-LANGU` contains the language selected at login

See also DELETE TEXTPOOL, READ TEXTPOOL, *Text elements*

Instance

There are three types of instances in an SAP system. The Application server instance, consisting of a dispatcher and all its work processes, the Message server, and the Gateway server. A typical system has a Message Server, a Gateway Server and one or more Application servers.

See also *Dispatcher*, *SAPSYSTEM, Work processes*

INT(x)

Use `TRUNC(x)` for the "integer" function.

Interactive Report

Description

Interactive reports are event-driven views of lists with drill-down capability to detailed lists. Interactive reports are limited to 9 levels of drill-down. The events you may use in these reports are `START-OF-SELECTION`, `GET...`, `END-OF-SELECTION`, `TOP-OF-PAGE...`, `AT PFnn`, `AT LINE-SELECTION`, `AT USER-COMMAND`. To return to the next higher list level, the user can press the "Back" (F3) function. To return to any higher level in code, assign the desired list level to `SY-LSIND`. The cursor-line string is in `SY-LISEL` upon valid selection. `HIDE` key fields to determine which record was selected.

System fields of interest to interactive reporting are:

Field	Description
`SY-TITLE`	Report title from report text elements or `SET TITLEBAR`
`SY-LINCT`	Number of lines in the list
`SY-LINSZ`	Line width in the list
`SY-PAGNO`	Current page number in the list
`SY-LINNO`	Current line number in the list
`SY-COLNO`	Current column number in the list
`SY-SROWS`	Number of lines in the window
`SY-SCOLS`	Number of columns in the window
`SY-CUROW`	Current row in the window
`SY-CUCOL`	Current column in the window
`SY-CPAGE`	Current page in the window
`SY-STARO`	Line number of the top displayed line
`SY-STACO`	Column number of leftmost displayed column
`SY-LSIND`	detail level currently being generated; basic list = 0
`SY-LISTI`	detail level selected
`SY-LILLI`	line number of selected line in window
`SY-LISEL`	contents of selected line

`INTERFACE`

Description

Defines an ABAP Objects interface.

This book doesn't discuss ABAP Objects programming.

See also *Object-oriented programming*

`INTERFACES`

Description

Includes the components and methods of the interface in an ABAP Objects class.

This book doesn't discuss ABAP Objects programming.

See also *Object-oriented programming*

Interrupt

There is no interrupt facility available to programmers in SAP except for program events and external events that can be triggered by using the "C" program SAPEVT.

See the Online Documentation CD for information following the path:

Basis Components
> *ABAP Workbench*
> *BC Basis Programming Interfaces*
> *Programming with the Background Processing System*
> *Using Events to Trigger Job Starts*
> *Triggering Events from External Programs*

`IS INITIAL`

Description

Relational condition: true when the subject of this clause equals its initial value. This condition may be used in **`CHECK`**, **`ELSEIF`**, **`IF`** and **`WHILE`**. It may not be used in **`WHERE`**.

Variable Type	Initial Value
C	(space)
N	(filled with zeros)
D	`'00000000'` (interpreted as 1/01/0001)
T	`'000000'` (interpreted as midnight)
I	0
P	0
F	0.0E+00
X	0

itab

Description

An internal (runtime) table that may contain query results, selection ranges or working data. Each `itab` has an `INDEX` which is a unique record number used in several commands. The "standard key" for an `itab` is the concatenation of all its fields which are not numeric (F, I, P) and not internal tables themselves. The standard key doesn't necessarily specify unique records. The commands and system fields that affect `itabs` include: `APPEND`, `CLEAR`, `COLLECT`, `DELETE`, `DESCRIBE`, `EDITOR CALL`, `FREE`, `INSERT`, `LOOP`, `MODIFY`, `PROVIDE`, `READ`, `REFRESH`, `SEARCH`, `SELECT`, `SORT`, `SPLIT`, `SUM`, `UPDATE`, `SY-TFILL`, `SY-TLENG`, `SY-TMAXL`, `SY-TNAME`, `SY-TOCCU`, `SY-TPAGI`, `SY-TTABC`, `SY-TTABI`

See also *Work Area* , [ ]

Language

See `SET LANGUAGE`

Launch a program

To execute an ABAP from your terminal use /SA38, /SE 38 or /SE37. To call another ABAP program or report from within an ABAP see `SUBMIT`. To launch an operating system (that is, UNIX or Windows NT) program on the application server from within an ABAP see `OPEN DATASET...FILTER`, and `/SM49`. To launch a program on the presentation server (local workstation) see `WS_EXECUTE`.

LDB

See *Logical Database*

LE

Relational operator "Less than or Equal".
See *Operators*

LEAVE

Syntax

```
LEAVE [PROGRAM].
```

Description

Forces an immediate return to the calling program from a report called with **SUBMIT...AND RETURN** or a transaction called with **CALL TRANSACTION** or **CALL DIALOG. LEAVE** has no effect in a report called by **SUBMIT** or a transaction called with **LEAVE TO TRANSACTION** or with a transaction code in the command field. The **IMPORT** objects are returned from the **CALL DIALOG. LEAVE PROGRAM** is identical to **LEAVE** except that it forces a return to the transaction selection screen from a report called by **SUBMIT** or a transaction called with **LEAVE TO TRANSACTION** or with a transaction code in the command field.

See also CHECK, CONTINUE, EXIT, REJECT, STOP

LEAVE LIST PROCESSING

Syntax

```
LEAVE LIST PROCESSING.
```

Description

Forces a return to the on-line transaction from the list-processing mode initiated by LEAVE TO LIST PROCESSING; processing continues with the PBO of the screen that controls the list. If the operator will interactively return to the on-line transaction with F3 (Back) or F15 (Exit) then this command is unnecessary.

LEAVE [TO] SCREEN

Syntax

```
LEAVE {SCREEN
| TO SCREEN scr}.
```

Description

Moves to the next default screen or to screen scr to continue processing in an on-line transaction. If scr = 0 then processing continues after the calling **CALL SCREEN** command.

See also SET SCREEN

LEAVE TO LIST PROCESSING

Syntax

```
LEAVE TO LIST PROCESSING [AND RETURN TO SCREEN scr].
```

Description

Branches from the on-line transaction to the list-processing mode. The list-processing mode completes with LEAVE LIST PROCESSING or with the operator selecting F3 (Back) or F15 (Exit). Upon return, processing continues with the PBO of the screen that controls the list, or optionally of screen `scr`.

LEAVE TO TRANSACTION

Syntax

LEAVE TO TRANSACTION `tcode` **[AND SKIP FIRST SCREEN].**

Description

Flushes all nested or stacked transaction calls, then calls transaction `tcode`. At completion, processing will continue with the original calling program. **AND SKIP FIRST SCREEN** will attempt to process the first screen in the background using SPA/GPA memory objects previously defined as PIDs or with `SET PARAMETER`. Since `SY-TCODE` contains the name of the current transaction, you can flush the stack of calls and re-start it with `LEAVE TO TRANSACTION SY-TCODE`.

See also `CALL TRANSACTION`

Line-break

See `NEW-LINE`

LOAD-OF-PROGRAM

Syntax

LOAD-OF-PROGRAM.

Description

(Release 4.6 and above) Event triggered one time immediately after a program of type 1, M, F, or S is loaded, before the selection screen of the program or its logical database is displayed. You may set the `PARAMETERS` and `SELECT-OPTIONS` defaults at this point. This will behave similar to the `INITIALIZATION` event.

See also *Events* for their triggering order

LOCAL

Syntax

LOCAL {f | <fs>}.

Description

Use in a **FORM...ENDFORM** subroutine to limit the scope of global field `f` or field-symbol `<fs>` to the current subroutine. Upon return from the subroutine, the values of the field or field-symbol are restored to their values before the subroutine was called.

See also `CONSTANTS`, `DATA`, `STATICS`, `TABLES`, `TYPES`

`LOG`

Syntax

```
LOG( y )
```

Description

Returns the logarithm base *e* of floating point number `y`, for `y > 0`.

See also *Arithmetic functions*

`LOG10`

Syntax

```
LOG10( y ).
```

Description

Returns the logarithm base 10 of floating point number `y`, for `y > 0`.

See also *Arithmetic functions*

Logical Database

A pre-defined hierarchical structure of database tables, linked with foreign keys for reporting purposes, and an associated ABAP program that reads all their records in hierarchical order. You may create and display Logical Databases in /SLDB or /SE36.

See *Appendix G.3*, `GET`, `REPORT ... DEFINING LDB`

Logical Expressions

See *Condition*, *Operators*

`LOOP`

Syntax

```
LOOP ... ENDLOOP.
```

Description

Analyzes the data set of extracted field groups. You declare the field group using `FIELD-GROUPS`, define the field group by inserting field(s) from one or more tables, and fill the field group from the tables using `EXTRACT`. Records defined by different field groups can be interleaved. These interleaved records can be sorted and then analyzed in a `LOOP ... ENDLOOP` code block.

`SY-SUBRC`	0 after **`ENDLOOP`** if a loop pass was executed
	4 otherwise

See also `CNT`, `EXTRACT`, `FIELD-GROUPS`, `INSERT`, `ON CHANGE OF`, `SUM`

LOOP AT ITAB

Syntax

```
LOOP AT itab
[ INTO wa
| TRANSPORTING NO FIELDS]
[FROM ndx1]
[TO ndx2]
[WHERE condition].
 ...
ENDLOOP.
```

Description

Loops through `itab` and fills the header line (or optionally, work area `wa`) with the contents of the current record then executes the commands before the **ENDLOOP**. If `itab` has no header line then you must **LOOP INTO** `wa`. **TRANSPORTING NO FIELDS** doesn't fill the header line or `wa`; use this addition to count the number of found records. **FROM** starts from record number `ndx1` where `ndx1 > 0`. **TO** continues to and includes record number `ndx2` where `ndx2 ≥ ndx1`. The comparisons in `condition` must start with a field in the structure of `itab`. `CONTINUE` unconditionally jumps up to **LOOP** for the next iteration. `CHECK <condition>` jumps up to **LOOP** for the next iteration if the condition is false, and continues execution with the next command if it's true. **EXIT** immediately terminates the loop. These conditions are true in loops if the **FROM**, **TO** or **WHERE** options are not used: `AT FIRST...ENDAT.` during the first iteration
`AT NEW f1...ENDAT.` when `f1` has just changed *
`AT END OF f1...ENDAT.` when `f1` will change at the next pass *
`AT LAST...ENDAT.` during the last iteration
* (or one of the fields to the left of `f1`)
`SY-TABIX` contains the current record number until the loop completes, then returns to its value before the **LOOP** command. `SY-TFILL` contains the current number of records in `itab` inside the loop and after **ENDLOOP**. `SY-TOCCU` contains the declared **OCCURS** value for `itab` inside the loop and after **ENDLOOP**.

`SY-SUBRC` 0 after **ENDLOOP** if any records were found
4 otherwise

See also APPEND, AT...ENDAT, COLLECT, DELETE, DO, INSERT, MODIFY, READ TABLE, SORT, WHILE

LOOP AT SCREEN

Syntax

```
LOOP AT SCREEN. ... ENDLOOP.
```

Description

While in a screen, the system itab `SCREEN` contains the following attributes for each field in the screen. **`LOOP AT SCREEN`** provides the ability to alter those attributes in the header line at runtime in the PBO module. You'll need to convey those changes to the screen using `MODIFY SCREEN`. In step-loop processing the change applies to the current line; in list processing it applies to the entire screen; in table-control processing you must make the change inside the loop (for each line). The structure of `SCREEN` is:

Field Name	Len	Type	Description
Screen-Name	30	C	Field name
Screen-Group1	3	C	Identifies field group 1
Screen-Group2	3	C	Identifies field group 2
Screen-Group3	3	C	Identifies field group 3
Screen-Group4	1	C	Identifies field group 4
Screen-Required	1	C	1 = input is mandatory
Screen-Input	1	C	1 = field can accept input
Screen-Output	1	C	1 = output will display
Screen-Intensified	1	C	1 = field is highlighted
Screen-Invisible	1	C	1 = field is invisible
Screen-Length	1	X	Field length
Screen-Active	1	C	① 1 = field is active
Screen-Display_3D	1	C	1 = allows I/O 3-D frame
Screen-Value_Help	1	C	1 = includes help button
Screen-Request	1	C	1 = Triggers the ON-REQUEST event in PAI

① `Screen-Active = 0` forces `Input = 0`, `Output = 0`, and `Invisible = 1`

Lower case

See `TRANSLATE`

LT

Relational operator "Less Than".

See *Operators*

Macros

See DEFINE

Memory

See EXPORT, IMPORT, *SPA/GPA*

MESSAGE

Syntax

```
MESSAGE {tnnn[(msgid)] | ID msgid TYPE t NUMBER nnn}
[WITH f1 [f2 [f3[ f4]]]]
[RAISING exception1].
```

Description

Sends message type t number nnn from message-id group msgid or the one shown in the **REPORT** statement.

Type	Name	Description
A	Abend (abnormal end)	Operator must restart transaction.
E	Error	Operator must enter correct data and press Enter.
I	Information	Operator must press <Enter> to continue.
S	Success	Notification on following screen.
W	Warning	Operator must enter new correct data or press <Enter>.
X	Exit	The transaction is terminated with a short dump.

WITH f1... inserts up to four parameters (up to 50 characters each) in the message consecutively at the positions of the & characters in the message text. **RAISING** triggers the exception within a function module. The text of messages is stored in table T100. The expanded message text is displayes in the status bar or in a modal dialog box. After a **MESSAGE** command completes, the seven system fields shown below are assigned. Use the following syntax to issue the online message returned from CALL TRANSACTION.

```
MESSAGE ID SY-MSGID TYPE SY-MSGTY NUMBER SY-MSGNO
  WITH SY-MSGV1 SY-MSGV2 SY-MSGV3 SY-MSGV4.
```

Message server

An instance of the SAP system.

See also *Instance*

METHOD

Description

Implements a method in an ABAP Objects class.

This book doesn't discuss ABAP Objects programming.

See also *Object-oriented programming*

METHODS

Description

Declares ABAP Objects methods in class and interfaces.

This book doesn't discuss ABAP Objects programming.

See also *Object-oriented programming*

MOD

Modulo operator (remainder from integer division).

See *Operators, arithmetic*

MODIFY DBTAB

Syntax

```
MODIFY [CLIENT SPECIFIED] dbtab….
```

Description

Executes `UPDATE` if the key is already present in `dbtab`, and executes `INSERT` otherwise. The command normally affects only the current client records. If you use **`CLIENT SPECIFIED`** then client (MANDT) becomes a normal field and you can affect other client's records. **`MODIFY`** is resource-intensive; use `UPDATE` or `INSERT` if you can determine which applies. The syntax is identical to that of `UPDATE` and `INSERT`.

See also `DELETE`, `INSERT`, `UPDATE`

MODIFY itab

Syntax

```
MODIFY itab
[FROM wa]
[INDEX ndx | TRANSPORTING f1...[WHERE <condition>]].
```

Description

Overwrites the current record of `itab` from the header line or optionally from work area `wa`. **`INDEX`** overwrites record `ndx` of `itab`; if record `ndx` is not present in `itab` then no action takes place. **`TRANSPORTING`** modifies only the listed fields, and **`WHERE`** modifies them only in records that match the condition.

<u>`SY-SUBRC`</u> 0 successful (**`INDEX`** only)

4 **`INDEX`** too large; no change

See also APPEND, INSERT

MODIFY ... LINE

Syntax

```
MODIFY
{ CURRENT LINE | LINE n1
[ OF CURRENT PAGE | OF PAGE p1 | INDEX ndx ] }
[ FIELD VALUE f1 [FROM g1] [f2 [FROM g2...] ] ]
[ FIELD FORMAT f1 f_format1 [f2 f_format2...] ]
[ LINE FORMAT l_format3 ].
```

Description

Modifies a line in the list, then writes the revised string back to that line. **CURRENT LINE** modifies the current line. **LINE** modifies line n1 in the specified the page. **INDEX** modifies the selected line on the list level identified by that index. List value f1 is changed to the value of the variable f1 or g1. Field and line attributes are set by f_format and l_format which are described in FORMAT.

SY-SUBRC	0 the line exists > 0 otherwise
SY-LISEL	contents of the last-read line which was modified
SY-LISTI	index of the of the last read
SY-LILLI	line number last read
SY-CPAGE	page number of the last read
SY-LSIND	index of the current screen

MODULE

Syntax

```
MODULE mod1
[ON [CHAIN] INPUT
|ON [CHAIN] REQUEST
|AT EXIT-COMMAND
|AT CURSOR-SELECTION].
```

Description

For use in module pools only. In screen flow control, this command calls module mod1 located in the associated module pool. **ON INPUT** calls the module only if the triggering field has a non-initial value. **ON REQUEST** calls the module only if the triggering field has a new value. **ON CHAIN INPUT** calls the module only if any field in the chain has a non-initial value. **ON CHAIN REQUEST** calls the module only if any field in the chain has a new value. **AT EXIT-COMMAND** calls the module before the input checks if the user triggered a type E function. **AT CURSOR-SELECTION** calls the module if the user triggered a type S function CS (typically F2).

See also *Flow Control*, LEAVE [TO] LIST PROCESSING

MODULE ... ENDMODULE

Syntax

`MODULE mod1 {OUTPUT | [INPUT]}...ENDMODULE.`

Description

For use in module pools only. A code block containing ABAP code that handles online processing events from screen flow control. **OUTPUT** specifies a module that will called from a PBO. **INPUT** specifies a module that will be called from a PAI; this is the default so you may leave it off.

MOVE

Syntax

`MOVE a TO b [PERCENTAGE n [RIGHT]].`

or

`MOVE a[+p1][(w1)] TO b[+p2][(w2)].`

Description

Assignment command, equivalent to `b = a`. The offset and width parameters are equivalent to: `b+p2(w2) = a+p1(w1)`, except in **MOVE** the offsets and widths may be variables; see String Handling. **MOVE** converts unlike fields; see "Type Conversions" for conversion information.

Example

```
p1 = 5.   w1 = 3.
p2 = 2.   w2 = 1.
s1 = '1234567890'.
S2 = 'ABCDEFGHIJ'.
MOVE s2+2(1) TO s1+5(3). → s1 contains '12345C 90'
MOVE s2+p2(W2) TO s1+p1(W1). → s1 contains '12345C 90'
```

PERCENTAGE transfers the left-most `n` percent of the <u>declared</u> length of `source` to `target`, and left-justifies it in `target` unless **RIGHT** is specified. `a` and `b` must be type `C` fields and `n` must be a field with a numeric value between `0` and `100`. If $n < 0$, the command will use 0, if $n > 100$, it will use 100. Use this for displaying "progress thermometers", etc.

Example

```
DATA: cvar1(100), cvar2(50), cpct(3).
PARAMETERS: percent(4) TYPE I DEFAULT 25.
Cpct      = percent.
cpct+2    = '%'.
cvar1     = '...1...2...3...4...5...6...7...8...9...0'.
MOVE cvar1 TO cvar2 PERCENTAGE percent.
WRITE: / cpct, cvar2.
```

Results in

```
25% ...1...2...3...4...5
```

Notice that it assigned `25` percent of the declared length `(100)`, not of the actual length `(50)`.

See also =, MOVE-CORRESPONDING, WRITE TO

MOVE-CORRESPONDING

Syntax

MOVE-CORRESPONDING array1 **TO** array2.

Description

If array1 and array2 are structured work areas such as header lines, then this command assigns values between like-named fields from array1 to array2

Example

```
MOVE array1-field1 TO array2-field1.
MOVE array1-field2 TO array2-field2.
etc.
```

MULTIPLY

Syntax

MULTIPLY a **BY** b.

Description

Equivalent to: `a = a * b.` Non-numeric fields are converted; see "Type Conversions" for conversion information.

See also ADD, DIVIDE, SUBTRACT

MULTIPLY-CORRESPONDING

Syntax

MULTIPLY-CORRESPONDING array1 **BY** array2.

Description

If array1 and array2 are structured work areas such as header lines, then this command multiplies like-named fields in array1 and array2.

Example

```
MULTIPLY array1-field1 BY array2-field1.
MULTIPLY array1-field2 BY array2-field2.
etc.
```

See also ADD-CORRESPONDING, DIVIDE-CORRESPONDING, SUBTRACT-CORRESPONDING

Names

Description

Field names are limited to 30 characters and must include at least one non-number character. They may not include spaces, parentheses, plus (+), minus or hyphen (-), commas or periods. Reserved words may not be used for field names. ABAP keywords are all reserved. `INITIAL` and `SPACE` are reserved words. You should consider all the words that appear as options ("additions") in commands to be reserved, such as `EXPORTING`, `CHANGING`, `INDEX`, etc. Program names are limited to 8 characters up to release 3.1 (30 characters later) which may be letters, numbers and the underscore. SAP requires that the name for a user-created program begin with "Y" or "Z". SAP suggests you assign the remaining letters as follows:

First letter indicates a user program: Y or Z

Second letter indicating the SAP application, such as

- A AM - Assets Management
- C PP - Production Planning
- E EDI
- F FI - Financial Accounting
- G GL - General Ledger
- I PM - Plant Management
- K CO - Controlling
- M MM - Materials Management
- P PS - Project System
- Q QM - Quality Management
- R EIS - Executive Information System
- S Basis
- U Utility
- V SD - Sales & Distribution
- Y System

Third Letter indicating the type of program

- F Function module code
- I Include module
- P Report programs
- V Update report

Use letters 4 and beyond for a string that may have meaning to the program. Leave at least the last character empty so it's available for future revisions.

Names of other objects vary in length and construction. SAP requires that the name for many user-created objects must begin with "Z" or "Z_" if you intend to move the program to production.

NB

Relational operator "Not Between" used in "ranges" table.
See also RANGES, SELECT-OPTIONS

NE

Relational operator "Not Equal".
See *Operators*

NEW-LINE

Syntax

```
NEW-LINE [[NO] SCROLLING].
```

Description

Creates a new line in a report. **NEW-LINE** is ignored if the current line is empty (it won't skip blank lines). **NO SCROLLING** locks subsequent lines from horizontal scrolling, creating title lines. **SCROLLING** clears the horizontal scroll lock (default).
See also SET LEFT SCROLL-BOUNDARY

NEW-PAGE

Syntax

```
NEW-PAGE
[NO-TITLE | WITH-TITLE]
[NO-HEADING | WITH-HEADING]
[PRINT ON | PRINT OFF]
[LINE-COUNT lin]
[LINE-SIZE col].
```

Description

Starts a new page in a report. **NEW-PAGE** is ignored if the current page is empty (it won't create blank pages). It triggers the **TOP-OF-PAGE** event but not the **END-OF-PAGE** event. SY-PAGNO contains the current page number. **NO-TITLE** ceases printing title, date and page number (default on detail lists). **WITH-TITLE** starts printing title, data and page number (default for basic lists). **NO-HEADING** ceases printing column headings (default on detail lists). **WITH-HEADING** starts printing column headings (default on basic lists). **PRINT ON** sends subsequent list output to printer vs screen. **PRINT OFF** restores output to screen. **LINE-COUNT** sets new number of lines per page. Set lin to 0 for unlimited length. The default is the value in REPORT. **LINE-SIZE** sets new number of columns per line and is effective only at the top of a list level (before the first WRITE or SKIP). The default is the value in REPORT. Set col to 0 for standard window width.
SY-PAGNO current page number

NP

String comparison operator "Not contains Pattern".
See *Operators*

Notation, York-Mills

See *Appendix I*.

Object Linking and Enabling

See *OLE*

Object names

See *Names*

Object–oriented programming

This book doesn't address ABAP Object programming. Commands that are strictly limited to Object-oriented programming include `ALIASES`, `CALL METHOD`, `CLASS`, `CLASS-DATA`, `CLASS-METHODS`, `CLASS-EVENTS`, `CREATE OBJECT`, `EVENTS`, `INTERFACE`, `INTERFACES`, `METHOD`, `METHODS`, `PRIVATE`, `PROTECTED`, `PUBLIC`, `RAISE EVENT`, `SET HANDLER`.

OCCURS

The `occurs` value of `itab`. The system field SY-TOCCU also contains the number after this command is executed. Note: SAP may change the `occurs` value from the `DATA` value during execution.
See `DATA`, `DESCRIBE TABLE`

okcodes

Description

In screen processing, okcode is the name of an event triggered by the user, or issued in **`SET USER-COMMAND`**. In BDC sessions it is the name of an event forced by the session. Some examples are:

okcode	Use	Description
`/nn`	③	Function Key nn
`/0` (zero)	③	<Enter> (This works but I've found no documentation for it)
`/8`	③	`F8`; Continue or Execute
`/11`	③	`F11`; Post
`%EX`	①②	Exit - Depart this process (yellow up arrow)
`%SC`	①	Display the dialog box "`Find by...`"

okcode	Use	Description
BACK	①③	F3; return to previous screen, ignoring required entries (same as EXIT)
CS	①③	F2; Cursor Select; double-click; (replaces PICK)
DLT	①②③	F14; Delete
EXIT	①③	F15; return to previous screen, ignoring required entries (same as BACK)
FCnn	②	"Function Key" nn created by SELECTION-SCREEN...FUNCTION KEY
HELP	②	F1; show the help screen for the current field
LIST	②③	F4; List the possible entries for the current field
MENU	②③	F10; move focus to the menu
P–	①③	F21; Scroll up to top of list (use SCROLL in screen processing)
P-	①③	F22; Scroll page up
P+	①③	F23; Scroll page down
P++	①③	F24; Scroll down to end of list
PFnn	②	Function Key nn
PICK	①③	F2; select; double-click; (replaced by CS)
PRI	①③	F13; Print
RW	①③	F12; Cancel (that is, Rollback Work)
SAVE	②③	F11; Save
tcode	③	call transaction tcode

① Some commands are trapped and processed by SAP, so they aren't available in list processing.
② User commands defined in the menu painter are available in SY-UCOMM for processing in AT USER-COMMAND and AT LINE-SELECTION.
③ Most okcodes can be used in BDC sessions.

See also *Appendix G.1*, AT USER-COMMAND, AT LINE-SELECTION, SET USER-COMMAND

OLE (Object Linking and Enabling)

Description

The Microsoft protocol for connecting data between applications. See these ABAP OLE commands: CALL METHOD, CREATE OBJECT, FREE OBJECT, GET PROPERTY, SET PROPERTY.

ON CHANGE OF

Syntax

```
ON CHANGE OF f1 [OR f2 [OR f5...]]...ENDON.
```

Description

Control block used in `GET` events and `SELECT...ENDSELECT` loops that is executed in the first record and whenever any of the listed fields change. It is not executed in the last record unless its value just changed.

See also `AT END OF...`, `AT NEW...`

OPEN CURSOR

Syntax

```
OPEN CURSOR cname FOR SELECT (select parameters).
```

Description

Creates table cursor (record pointer) `cname` in the table generated by the imbedded **`SELECT`** command and initially locates it in the first record of that table. Any `SELECT` command may be used that produces a table (that is, `SELECT SINGLE` and aggregates are not allowed). Records in that resultant table may be read with `FETCH`. `cname` must be `TYPE CURSOR`. Table cursors are closed by screen changes, `CLOSE CURSOR`, `COMMIT WORK`, *RFCs*, `ROLLBACK WORK`.

See also `FETCH`, `CLOSE CURSOR`

OPEN DATASET

Syntax

```
OPEN DATASET filename FOR {INPUT | OUTPUT | APPENDING}
IN {BINARY | TEXT} MODE
[AT POSITION <pos>]
[MESSAGE msg]
[FILTER cmd]
[TYPE attr].
```

Description

Opens an external file accessible from the application server. `filename` must follow the operating system conventions (such as lower case, forward slashes: `'/tmp/myfile'`, etc.). **`INPUT`** opens the file read-only (default). **`OUTPUT`** opens it in write mode, overwriting the existing same-named file. **`APPENDING`** writes to the end of the existing file. **`IN BINARY MODE`** (default) concatenates subsequent blocks with no delimiter. **`IN TEXT MODE`** terminates each block with the newline character. NOTE: trailing blanks are truncated when writing in text mode. To create fixed-length records use binary mode to write each record and append your own line break (CRLF = Hex `0D0A`). **`AT POSITION`** establishes the byte number for the next read or write. **`MESSAGE`** assigns to `msg` any operating system message generated by the command. **FILTER** cmd] Passes the command to the operating system. **FILTER** will pass the input or output file (as appropriate) through the operating system filter cmd (such as UNIX 'compress' for output files or 'uncompress' for input files). You can also use this option to launch an operating system program that is independent of data files - just use an empty field for filename, and send the entire command in cmd. **TYPE** passes the attribute string to the operating system.

Example

To rename a file in the UNIX operating system:

```
DATA: nullfile, unixcmd(80) VALUE 'mv '.
PARAMETERS: ORG_FILENAME(32), NEW_FILENAME(32).
unixcmd+4  = org_filename.
unixcmd+40 = new_filename.
CONDENSE unixcmd.
OPEN DATASET nullfile FILTER unixcmd.
```

READ DATASET will attempt to open an unopened file FOR INPUT IN BINARY MODE. TRANSFER will attempt to open an unopened file FOR OUTPUT IN BINARY MODE. Use the debugger for current file status: /H F3 {Goto {System {System areas <Area = 'Datasets'

SY-SUBRC 0 if successful
8 otherwise

See also CLOSE DATASET, DELETE DATASET, READ DATASET, TRANSFER

Open SQL

Description

A subset of SQL commands that are executed by SAP and provide the relational integrity and a safety net not available in native SQL. These commands include: CLOSE CURSOR, COMMIT WORK, DELETE, FETCH, INSERT, MODIFY, OPEN CURSOR, ROLLBACK WORK, SELECT, UPDATE.

See also EXEC SQL

Operators

Description

Note: All operators (including grouping parentheses) must be separated by spaces. For example: x = 1 + (a + (b * c)).

Arithmetic Operators

+ - * /

** (that is, exponentiation)

DIV integer division: a **DIV** b == **TRUNC**(a / b)

MOD modulo operator (remainder):

a **MOD** b == a - b * (a **DIV** b)

where 0 ≤ (a **MOD** b) < **ABS**(b)

Order of precedence for arithmetic operations (evaluated left-to-right, except exponentiation is evaluated right-to-left):

- grouping operators (parentheses)
- functions
- exponentiation (**)
- **DIV MOD * /**
- **+ -**

Bit Operators (Operands and result must be Type X)

BIT-AND Bit-by-bit logical product of the operands

BIT-NOT Bit-by-bit logical inverse of the operands

BIT-OR Bit-by-bit logical sum of the operands

BIT-XOR Bit-by-bit exclusive-or of the operands

Order of precedence for bit operations (evaluated left-to-right):

- grouping operators (parentheses)
- **BIT-NOT**
- **BIT-AND**
- **BIT-XOR**
- **BIT-OR**

Boolean Operators

OR AND NOT (there's no **XOR** in ABAP)

Relational Operators

= or **EQ**

>< or <> or **NE**

> or **GT**

>= or => or **GE**

< or **LT**

<= or =< or **LE**

BETWEEN v1 **AND** v2

IN rtab (rtab is an internal table as described in RANGES)

IS INITIAL (= 0 or ' ' (SPACE) depending on type)

Order of precedence for logic operations: (evaluated left-to-right until the result is determined, then evaluation stops)

- grouping operators (parentheses)
- **NOT**
- **AND**
- **OR**
- Relational Operators

See also *Condition*

String Comparison Operators

a **CA** b a Contains Any one or more characters from b

a **NA** b a does Not contain Any characters from b

a **CO** b a Contains Only characters from b

a **CN** b a Contains any characters Not from b

a **CS** b a Contains the String b
(case-insensitive, trailing blanks ignored)

a **NS** b a does Not contain the String b

a **CP** b a Contains the descriptive Pattern b

a **NP** b a does Not contain the descriptive Pattern b
Pattern elements are:
* in b represents any string
+ represents any single character
# forces the next character to be compared exactly
(upper & lower case, *, +, #, and trailing space)

SY-FDPOS is set to the zero-based offset in a of the first match after each successful string comparison, and either to the length of a or to the offset in a of the first mismatch after a failed comparison.

Example

'ABAP Commands' CS 'MA' → SY-FDPOS contains 8

'ABAP Commands' CS 'QQ' → SY-FDPOS contains 13

Bit Comparison Operators

Where b is type X, the comparison is true if between corresponding bit positions in a and b:

a **O** b every “1” in b is “1” in a

a **Z** b every “1” in b is “0” in a

a **M** b of the “1’s” in b at least one is “1” and at least one is “0” in a

See also *Condition*

Output length

Description

The field length as it would be printed by the `WRITE` command. For example Type C's length and output length are the same, while Type P's output length is twice its length.

See also `DESCRIBE FIELD`

OVERLAY

Syntax

OVERLAY s1 **WITH** s2 **[ONLY** s3**].**

Description

Replaces spaces in `s1` with the character in that position in `s2`, leaving `s2` unchanged. **ONLY** `s3` replaces the character in `s1` with the character in that position in `s2` if the `s1` character is any one of the characters in the `s3`, leaving `s2` and `s3` unchanged.

Example

```
DATA: name(24) VALUE 'ABAP Command Book',
      ul(24) VALUE '________________________',
      date(10) VALUE '07-04-2001',
      sep(10) VALUE '//////////'.
OVERLAY name WITH ul. → name contains 'ABAP_Command_Book'
OVERLAY date WITH sep ONLY '-.'. → date contains
'07/04/2001'
```

`SY-SUBRC` 0 if any character is replaced

See also the other string processing commands: `CONCATENATE`, `CONDENSE`, `REPLACE`, `SEARCH`, `SHIFT`, `SPLIT`, `STRLEN()`, `TRANSLATE`

Packed field

Description

A packed field stores two digits per byte in Binary-Coded Decimal(BCD) format, reserving the first nibble (half-byte) for the sign, so its resolution is twice its length minus 1.

See also *Type Conversion*, `UNPACK`

Page Break

See `NEW-PAGE`

Page length

See `NEW-PAGE`, `REPORT`

PAI

See `PROCESS AFTER INPUT`

Parameter ID

Description

Label for a default value stored in user's SPA/GPA memory area.

See GET PARAMETER ID, SET PARAMETER ID, *SPA/GPA memory area*

PARAMETERS

Syntax

```
PARAMETER[S] p1[(len)]
[DEFAULT f1]
[TYPE t1 [DECIMALS n] | LIKE v1]
[LOWER CASE]
[OBLIGATORY]
[MEMORY ID pid1]
[MATCHCODE OBJECT mc1]
[MODIF ID mid1]
[NO DISPLAY]
[AS CHECKBOX]
[RADIOBUTTON GROUP g1 [DEFAULT 'X']].
```

Description

Defines input parameters on a selection screen for a report. Parameter names are limited to eight characters in length. **DEFAULT** populates the field with f1 and offers that default value to the user. **TYPE** and **LIKE** declare the type of the parameter. **DECIMALS** only applies to Type P. **LOWER CASE** allows upper & lower case (that is, for Unix file names). **MEMORY ID** defaults the parameter to the value of pid1. **MATCHCODE OBJECT** assigns matchcode object mc1 to the parameter. **MODIF ID** assigns screen modification group id mid1 to the parameter (see SCREEN-GROUP1 in LOOP AT SCREEN). **NO DISPLAY** hides the parameter; you may extract its value with LOOP AT SCREEN. **AS CHECKBOX** is always Type C(1) and equals either 'X' (Yes) or SPACE (No). The **CHECKBOX** and **RADIOBUTTON** 'X' must be capitalized. To use radiobuttons, you must have at least two **PARAMETERS** commands with the same radiobutton group. When the selection screen opens, the default radiobutton will be "pressed". If the user presses another, then all other radiobuttons are cleared; only one radiobutton at a time may be pressed. The first radiobutton in a group is the default unless you declare a **DEFAULT**. You must individually test each radiobutton **PARAMETERS** command in a group; there's no way to test the group.

See AT SELECTION-SCREEN, SELECTION-SCREEN, SELECT-OPTIONS

Pattern Characters (Wildcards)

Description

Wildcards in SAP depend on the environment:

Environment	Single Char .	String
Repository Info Sys (SE85)	+	*
string comparisons	+	*
WHERE ... LIKE	_	%
Illegal Password table USR40	?	*
Screen Painter (see Note)	_	(na)

Note: Place a "v" at the end of a screen painter field to provide for negative numbers. Place a comma or period in the screen painter field to force a decimal point in that position.

Pause

Use the Function Module `RZL_SLEEP` to pause the program a defined number of seconds. Hint: to insert function module calls into your program, use [Pattern.

PBO

See `PROCESS BEFORE OUTPUT`

Percentage

See `MOVE ... PERCENTAGE`

PERFORM

Syntax

```
PERFORM formname
[(ext_prog_name) | IN PROGRAM ext_prog_name [IF FOUND]]
TABLES itab1 itab2
USING a1 a2
CHANGING a3.
```

Description

Calls a subroutine created by the FORM...ENDFORM statements in the current report or in the external report or program ext_prog_name. Its actual parameters must have matching formal parameters in the called FORM statement. Parameters are positional and any number of parameters may be included. The **CHANGING** attribute must match the **CHANGING VALUE** attribute in the called FORM statement.

IN PROGRAM ext_prog_name **IF FOUND** calls the subroutine in the external program if the subroutine exists, otherwise it continues with the next statement. NOTE: this is an undocumented option and may not apply to all releases of SAP. **IN PROGRAM** (SY-REPID) **IF FOUND** calls the subroutine in the current program if the subroutine exists, otherwise it continues with the next statement.

You can issue multiple calls with the colon-and-comma construction:

```
PERFORM formname: USING a11 a12, a21 a22, a31 a32,....
```

Pf-status

Description

In SAP a "status" or "pf-status" is a transaction screen. Screens are created in the screen painter SE51, and their menus in the menu painter SE41.

See also SET PF-STATUS

PID

See Parameter ID

POH

See PROCESS ON HELP-REQUEST

POSITION

Syntax

```
POSITION c1.
```

Description

The next WRITE statement will begin in column c1 and overwrite anything previously written from column c1.

See also WRITE

`POV`

See `PROCESS ON VALUE`

Presentation server

Description

The top layer of SAP's three-layer Client-Server-Server architecture. This is the user's workstation, running the SAPGUI, and could be referred to as the presentation client.

See also *Application Server, Database Server, SAPGUI*

`PRINT-CONTROL`

Syntax

```
PRINT-CONTROL
[CPI w1]
[LPI h1]
[SIZE s1]
[COLOR { BLACK | RED | BLUE | GREEN | YELLOW | PINK }]
[LEFT MARGIN col1]
[FONT f1]
[FUNCTION f2]
[LINE lin1]
[POSITION col2].
```

Description

Sets the format of subsequent printer output from the current location, or from the location specified by either or both of **`LINE`** and **`POSITION`**. **`LINE`** and **`POSITION`** may only be used with one or more of the other options. Those options are: **`CPI`** - characters per inch, **`LPI`** - lines per inch, **`SIZE`** - size of typeface, **`COLOR`** - color of output, **`LEFT MARGIN`** - left margin in columns, **`FONT`** - name of font, **`FUNCTION`** - sub-argument as found in table `T022D`. **`PRINT-CONTROL`** affects printer output only; use `FORMAT` to set screen attributes.

This command uses the contents of tables `TSP03` and `T022D` to determine the printer codes required to execute this format; if the format isn't supported by the printer, this command is ignored. `TSP03` maps from the Output Device of the assigned printer to its Device Type. `T022D` maps from the Device Type and **`PRINT-CONTROL`** to the printer-specific control characters. Those control characters are the escape commands needed to produce the requested output.

See also `FORMAT`, `WRITE`

PRIVATE

Description

Defines the private section of an ABAP Objects class.
This book doesn't discuss ABAP Objects programming.
See also *Object-oriented programming*

PROCESS

Syntax

```
PROCESS
{ AFTER INPUT | BEFORE OUTPUT
| ON HELP-REQUEST | ON VALUE-REQUEST }.
```

Description

For use in module pools only. These are events in screen processing, triggered as follows. **AFTER INPUT** - the operator pressed a function key or a button, or selected a menu item. **BEFORE OUTPUT** - before the screen is displayed. Use this to initialize the field values and attributes. **ON HELP-REQUEST** - the operator selected Help (that is, F1). **ON VALUE-REQUEST** - the operator selected the Possible Entries list (that is, F4).
See also *Events, Flow Control*

PROGRAM

Syntax

```
PROGRAM progname.
```

Description

The header line in module pool source code – equivalent to REPORT.

Progress Thermometer

See MOVE ... PERCENTAGE

Property

Description

An attribute of an OLE object.
See GET PROPERTY, *OLE*, SET PROPERTY

PROTECTED

Description

Defines the protected section of an ABAP Objects class.
This book doesn't discuss ABAP Objects programming.
See also *Object-oriented programming*

PROVIDE

Syntax

```
PROVIDE
{f1 [f2 ...] | *} FROM itab1
{g1 [g2 ...] | *} FROM itab2
[h1 [h2 ...] | * FROM itab3 ...]
BETWEEN h1 AND h2.
   ...
ENDPROVIDE.
```

Description

Loops through `itab1`, `itab2` ... a number of times to retrieve a specific data series; this command is used for HR processing of Infotypes and has no other use in ABAP. HR programming is a specialty area that requires its own techniques.

See also `INFOTYPES`

PUBLIC

Description

Defines the public section of an ABAP Objects class.
This book doesn't discuss ABAP Objects programming.

See also *Object-oriented programming*

RAISE

Syntax

```
RAISE e1.
```

Description

In a function module, **`RAISE`** triggers the exception `e1`. If the calling program lists `e1` in its `EXCEPTIONS` list, then processing returns to the caller without assigning `EXPORT` values; otherwise the program terminates with an error message.

See also `MESSAGE...RAISING`

RAISE EVENT

Description

Triggers an ABAP Objects event.
This book doesn't discuss ABAP Objects programming.

See also *Object-oriented programming*

Random number

You can generate pseudo-random numbers ABAP by using the functions in program SAPLF052 (function group F052). Use transaction /SE37 {Function module F4 {Information system Function group = 'F052' to examine the available functions.

RANGES

Syntax

RANGES rtab **FOR** f1.

Description

Creates the internal table rtab for the field f1 with the structure:

Field Name	Type	Description
Sign	Type C (1)	{I (include) | E (exclude)}
Option	Type C (2)	Relational operator
Low	LIKE f1	Comparison value and inclusive low value for BT and NB
High	LIKE f1	Inclusive high value for BT and NB

which the program can then populate with conditions to be used in an IN condition. Multiple records are interpreted as OR alternatives. Supported relational operators are: EQ, NE, GT, GE, LT, LE, BT (BeTween), NB (Not Between), CP (Contains Pattern), NP (Not contains Pattern). For a pattern (in the Low field) use "+" for single characters and "*" for any number of characters. A range table is automatically created by SELECT-OPTIONS. You may use the condition IN rtab in CHECK, IF, SELECT, SUBMIT and WHERE statements.

READ DATASET

Syntax

READ DATASET filename **INTO** array1 **[LENGTH** w**] .**

Description

Reads the next record from the sequential file filename on the application server into the named structure. Attempts to OPEN the unopened file IN BINARY MODE FOR INPUT if the file is not already open. If the file is opened IN BINARY MODE then **READ DATASET** reads the number of bytes in array1. If the file is opened IN TEXT MODE then **READ DATASET** reads a line (that is, to the next newline character). [returns in [the number of characters read. The file contents are read literally and without conversion.

SY-SUBRC 0 Record was read successfully
4 End of file
8 Cannot open file

See also CLOSE DATASET, DELETE DATASET, OPEN DATASET, TRANSFER

READ ... LINE

Syntax

```
READ {CURRENT LINE | LINE n
[OF CURRENT PAGE | OF PAGE p]
[INDEX ndx1] }
[FIELD VALUE f1 [INTO g1]].
```

Description

Reads into SY-LISEL the contents of the current line or of line n of the current list, and refreshes all the hidden values. **OF CURRENT PAGE** reads line n on the current page of the list. **OF PAGE** reads line n on page p of the list. **INDEX** reads the line at list level ndx1 (not appropriate for **READ CURRENT LINE**). **FIELD VALUE** assigns the value of list element f1 (typically an input field, such as a checkbox) to the variable f1; **INTO** g1 assigns it into a variable having a different name. Remember to CLEAR f1 or g1 before each iteration.

SY-SUBRC 0 a line was read
>0 otherwise

SY-LISEL contains the last-read line

SY-LISTI contains the index of the of the last read

SY-LILLI contains the line number last read

SY-CPAGE contains the page number of the last read

SY-LSIND contains the index of the current screen

See MODIFY LINE, WRITE.

READ REPORT

Syntax

```
READ REPORT rpt1 INTO itab.
```

Description

Reads program rpt1 into itab. itab must be at least 72 characters wide.

SY-SUBRC 0 a line was read
>0 otherwise

See also DELETE REPORT, INSERT REPORT, READ TEXTPOOL

READ TABLE

Syntax

READ TABLE itab **[INTO** wa**]**
[INDEX ndx1 | **WITH KEY** {f1=v1 [f2=v2...] | k1 | =k2}
[BINARY SEARCH]]
[COMPARING f1 [f2...] | **[ALL FIELDS]]**
[TRANSPORTING f1 [f2...] | **[NO FIELDS]].**

Description

Reads into the header line or work area wa the first entry in itab whose values match the non-numeric, non-initial fields in the header line or wa.
INDEX reads the entry at record number ndx1.
WITH KEY... reads the first entry:

- whose fields match the variables v1, v2... (vn converted to match fn)
- whose table line matches the string k1 (itab converted to match k1)
- whose table line matches the string k2 (k2 converted to match itab).

You must SORT itab ascending (by f2 f2... in the case of that **KEY** option) before using **BINARY SEARCH**. **COMPARING...** compares the listed fields or all fields with the header line or wa after the line is read, and sets SY-SUBRC.
TRANSPORTING... assigns just the listed fields or no fields into the header line or wa; use **NO FIELDS** to just set SY-SUBRC and SY-TABIX.

SY-SUBRC 0 the read (and the compare if used) was successful
2 the read succeeded and the compare failed
4 the read failed

SY-TABIX record number that was read if successful

READ TEXTPOOL

Syntax

READ TEXTPOOL `rpt1` **INTO** `itab` **LANGUAGE** `lng1`.

Description

Reads from the library into `itab` the text elements for program `rpt1` in language `lng1`. You can find the login language in the system field SY-LANGU. The structure of the TEXTPOOL `itab` is:

Structure of TEXTPOOL

Fieldname	Description	Type
ID	ABAP textpool ID	C1
Key	Text element key	C8
Entry	Language-dependent text	C255
Length	Number bytes reserved for text	I

Types of text in TEXTPOOL

Application	ID	Key
Column heading	H	001-004 (line number)
Text symbols	I	NNN
Report or Program titles	R	(blank)
List headings	T	(blank)
Selection texts	S	Name up to 8 char

SY-SUBRC 0 if the textpool was read
non-zero otherwise

See also DELETE TEXTPOOL, INSERT TEXTPOOL, *Text elements*

RECEIVE RESULTS

Syntax

```
RECEIVE RESULTS FROM FUNCTION fnc1
[IMPORTING p1=f1 [p2=f2...]]
[TABLES p1=itab1 [p2=itab2...]]
[EXCEPTIONS e1[=retcode1] [e2[=retcode2...]]].
```

Description

Receives the results of an asynchronous function (that is, RFC) that was called using

```
CALL FUNCTION fnc1...
   STARTING NEW TASK taskname
   PERFORMING formname ON END OF TASK.
```

The **RECEIVE RESULTS** command must be in the `FORM formname` subroutine that's called by the function calling statement. The parameters for the **RECEIVE RESULTS** command must complement those of the function whose results it is receiving, that is, **IMPORTING** parameters here must match `EXPORTING` parameters in the function, **TABLES** must match `TABLES`, and **EXCEPTIONS** must match `EXCEPTIONS`. The `FORM` subroutine must have a `USING taskname` parameter in its definition.

`SY-SUBRC` 0 no exception raised
`retcode_n` if an exception was raised
> 0 otherwise

See also `CALL FUNCTION`

REFRESH CONTROL

Syntax

```
REFRESH CONTROL ctrl1 FROM SCREEN scr1.
```

Description

Restores `ctrl1` (created by a `CONTROLS` statement) to the initial state defined in `scr1`. Screen `scr1` need not be the same screen used in the [definition.

See also `CONTROLS`

REFRESH ITAB

Syntax

```
REFRESH itab.
```

Description

Empties `itab`. Allocated memory is not released. If there is a header line it remains unaffected.

See also `CLEAR`, `FREE`

REJECT

Syntax

`REJECT [dbtab].`

Description

Unconditionally jumps to the bottom of the current **GET** code block to get the next record of the current table. dbtab skips all **GET** events until the next record of dbtab is available. dbtab must be at the same or higher hierarchical level in the LDB as the current table. It must not be deeper in the hierarchy than the current table.

See also *Appendix G.3*, CHECK, CONTINUE, EXIT, LEAVE, STOP

Relational Operators

See *Operators*

Remote Function Call (RFC)

Description

Synchronous or asynchronous call to a function on a system other than that of the caller. The remote system may be another SAP product or a "C" program running on an application server.

See also CALL FUNCTION

REPLACE

Syntax

`REPLACE string1 [LENGTH w] WITH string2 INTO string3.`

Description

Searches string3 for the first (case-sensitive) occurrence of the first w characters of string1, removes from string3 those w characters and inserts at that location the entire contents of string2. string1 and string2 are unchanged.

Example

```
DATA city(20) VALUE 'Austin, TX'.
REPLACE 'TX' WITH 'Texas' INTO city.
```

city contains 'Austin, Texas'

SY-SUBRC	
0	if string1(w) was found in string3
> 0	otherwise

See also the other string processing commands: CONCATENATE, CONDENSE, OVERLAY, SEARCH, SHIFT, SPLIT, STRLEN(), TRANSLATE

REPORT

Syntax

REPORT rpt1
[DEFINING DATABASE ldb**]**
[LINE-SIZE n**]**
[LINE-COUNT l **[** (n) **]]**
[MESSAGE-ID xx**]**
[NO STANDARD PAGE HEADING] .

Description

Defines a report - the first line of the program. **DEFINING DATABASE** ldb specifies the use of the logical database ldb. **LINE-SIZE** defines the report line size ≤255; the default = 80. **LINE-COUNT** sets the number of lines per page in l, default = 0 (that is, no forced pagination), and sets the number of lines reserved for the end-of-page area in n, default = 0. **MESSAGE-ID** sets the ID values for online messages from table T100. **NO STANDARD PAGE HEADING** suppresses the title, date and page number header. Limit the name to 8 characters to release 3.1 (30 characters later) and don't start the name with "R".

SY-LINSZ contains the current line-size
SY-LINCT contains the current lines per page
SY-REPID contains the current report name

See also END-OF-PAGE, *Names*, PROGRAM, RESERVE

RESERVE

Syntax

RESERVE n **LINES.**

Description

In list generation, **RESERVE** tests whether at least n lines remain on the page. If not, then it triggers the END-OF-PAGE event, providing n Lines for end-of-page processing. n may be literal or a variable. I found this command very hard to get to work right, but REPORT rpt1 LINE-COUNT l(n) works exactly as advertised with no hassle.

See also BACK, END-OF-PAGE

RFC - Remote Function Call

See *Remote Function Call*

ROLLBACK WORK

Syntax

`ROLLBACK WORK.`

Description

Reverses all database changes made since the last **ROLLBACK WORK** or COMMIT WORK. Since this clears all table cursors, don't use it in SELECT...ENDSELECT loops or with FETCH commands. hassle. Using the **BYPASSING BUFFER** option in a **SELECT... ENDSELECT** loop posts changes directly to the database tables, so subsequent ROLLBACK WORK commands have no effect.

See also COMMIT WORK

Round

See TRUNC, WRITE ... ROUND

SAPGUI

Description

The presentation interface or the "client" in the SAP client-server-server architecture. The "top layer" of SAP, above the application server, residing on the user's terminal.

See also *Application Server, Database Server, Presentation Server*

SAPTEMU

Obsolete term for SAPGUI.

SAPSYSTEM

Description

All the instances plus the database make up a system. Available systems are listed in the table TSYST. Typical systems in a landscape are:

- Development (or Integration) System
- Test (or Quality Assurance or Consolidation) System
- Production (or Recipient) System

"System" in SAP is somewhat equivalent to "Instance" in Oracle.

See also *Instance*.

SAPSCRIPT

Description

SAPScript is a means for formatting reports for printed output.

See the on-line help [Basis Components [System Administration [Style and Layout Set Maintenance

Scope

`DATA`, `CONSTANTS` and `TABLES` declarations in a main program or its `TOP` include are global to that program, all its subroutines and all its `INCLUDEs`.

`DATA`, `CONSTANTS` and `TABLES` declarations in a subroutine are local to that subroutine and are initialized each call. These fields are not available to subroutines called from the one in which they were declared.

`STATICS` declarations work exactly like `DATA` inside a subroutine except the values of the variables thus declared are retained between calls to the subroutine.

`LOCAL` declarations use within a subroutine the attributes but not the values of global variables having the same names and restore their original values upon returning from the subroutine.

See also `FORM`

Screen

See `CALL SCREEN`, `LEAVE SCREEN`, *Pf-status*, `SET PF-STATUS`, `SET SCREEN`, `SET TITLEBAR`

Screen attributes table

See `LOOP AT SCREEN`.

SCROLL LIST

Syntax

```
SCROLL LIST
{ TO FIRST PAGE
| TO LAST PAGE
| TO PAGE p
| FORWARD [n PAGES]
| BACKWARD [n PAGES]
| TO COLUMN c
| LEFT [BY n PLACES]
| RIGHT [BY n PLACES] }
[ LINE r1 ]
[ INDEX ndx1 ].
```

Description

Scrolls the list in the current list level under program control and highlights the current line. **LINE** highlights line `r1`. **INDEX** jumps to list level `ndx1` then scrolls.

`SY-SUBRC` 0 successful
4 list area exhausted
8 list doesn't exist (index call)

`SY-LSIND` contains the current list level index

SEARCH

Syntax

```
SEARCH {string1 | itab} FOR string2
[ABBREVIATED]
[STARTING AT p1]
[ENDING AT p2]
[AND MARK].
```

Description

Searches `string1` or the table `itab` (not its header line) for the first case-insensitive match with the contents of `string2`. `string1` and `string2` are treated as character strings without conversion. `string2` may be of any of the forms:

String form	Description
`'abc '`	the string "abc" (dropping any trailing spaces)
`'.abc.'`	the string "abc" (including trailing spaces)
`'*abc'`	any string that ends in "abc"
`'abc*'`	any string that begins with "abc"

ABBREVIATED - `String2` may have missing characters after the first character. **STARTING** - the search starts the search at `string1` offset `p1` or `itab` record `p1`. **ENDING** - the search ends the search at `string1` offset `p2` or itab record p2. **AND MARK** - capitalizes the matched substring of string1. Substring delimiters for **AND MARK** are space, comma, semicolon, colon, period, plus, exclamation point, question mark, parentheses, forward slash and equals.

SY-SUBRC 0 if found
>0 otherwise

SY-FDPOS zero-based offset of the match in string1 or in line of itab

SY-TABIX record number in itab where the match was found

Example

```
DATA: s1(24) VALUE 'ABAP Computer Book'.
SEARCH s1 FOR 'cmptr' ABBREVIATED AND MARK.
WRITE: s1, SY-FDPOS. →ABAP COMPUTER Book 5
```

See also the other string processing commands: CONCATENATE, CONDENSE, OVERLAY, REPLACE, SHIFT, SPLIT, STRLEN(), TRANSLATE

SELECT COMMANDS

Description

These commands retrieve records from a database table or a view in a random order, and place them in the work area. Several forms are illustrated below. The order of the **FROM** and **INTO** parameters is not important.

SY-SUBRC 0 one or more records was retrieved
4 no match
8 key was incomplete (**SELECT SINGLE FOR UPDATE**)

SY-DBCNT current number of records retrieved,
and the total number retrieved after **ENDSELECT**

See also ON CHANGE OF..., WHERE

SELECT - BASIC FORM

Moves selected data from a database table or a view into the corresponding fields of the work area.

```
SELECT {f1 [AS g1] f2 [AS g3] f3 [AS g3]   | * | itab1 }
  FROM [dbtab | (dbtabname)]
  [CLIENT SPECIFIED]
  [INTO { [CORRESPONDING FIELDS OF] wa | AS (h1, h2, h3
  ...) } ]
  [UP TO n ROWS]
  [BYPASSING BUFFER]
  [WHERE <condition>]
  [ORDER BY {f1 f2 ... [DESCENDING]... | PRIMARY KEY |
  (itab2)}].
  ...
ENDSELECT.
```

Moves selected fields from one record at a time into the corresponding fields of the header line, or left-to-right into work area wa, or into **CORRESPONDING FIELDS OF** wa or left-to-right into the named fields h1, h2,... The selected fields may be all the fields in the table (*), explicitly listed fields (f1 f2 ...), or fields listed – one record at a time – in itab1. **AS** g1 points to field g1 in the target for **CORRESPONDING FIELDS** asignments. dbtab is the literal name of a table or view, and dbtabname is a variable whose value is the name of a table or view. **CLIENT SPECIFIED** turns off automatic client filtering so records for all clients or any specific client can be selected. **INTO** wa moves the retrieved data into work area wa. **UP TO** n **ROWS** stops selecting when n matching records have been found. **BYPASSING BUFFER** gets data directly from the database, not using buffered data. **WHERE** retrieves only records that satisfy the condition. <condition> may be any simple or compound logical expression on one or more of the field values. **See** Condition for details. Processing of **ORDER BY** precedes **UP TO** n **ROWS** if both options are used. If the **SELECT** statement has a field list (instead of "*") then the **ORDER BY** fields must appear in that list. You may use **ORDER BY PRIMARY KEY** only when you select all fields with "*". **ORDER BY** (itab2) works the same as by f1 f2 ... if the itab has one type C field up to 72 characters long, and contains one column name per record. Table cursors are cleared by COMMIT WORK and ROLLBACK WORK so any commands that invoke them shouldn't be used in a **SELECT... ENDSELECT** loop or before a FETCH command. Those commands include: CALL SCREEN, CALL DIALOG, CALL TRANSACTION, MESSAGE, BREAKPOINT.

SELECT - COLUMNS

```
DATA: w1 LIKE dbtab-f1 OCCURS 5, w2 LIKE dbtab-f2
OCCURS 5.
SELECT [DISTINCT] f1 f2 FROM [dbtab | (dbtabname)]
  INTO (w1, w2)
  [UP TO n ROWS]
  [BYPASSING BUFFER]
  [WHERE...]
  [ORDER BY...].
  ...
ENDSELECT.
```

Retrieves selected fields into internal tables. **DISTINCT** retrieves unique combinations of values of the selected fields. **BYPASSING BUFFER** is unnecessary with **DISTINCT** and **ORDER BY**. The data are available after **ENDSELECT** in the internal tables.

SELECT - SINGLE

```
SELECT SINGLE [FOR UPDATE] {f1 f2 f3...| *}
  FROM [dbtab | (dbtabname)]
  [INTO wa]
  [UP TO n ROWS]
  WHERE key1 = v1 [AND key2 = v2..].
```

Retrieves the first matching record, or retrieves the single matching record if the entire key value is specified. Notice: no **ENDSELECT** is needed. **FOR UPDATE** protects the selected record from changes by others until the next COMMIT WORK. Use SELECT...UP TO 1 ROWS rather than **SELECT SINGLE** if you can't fully qualify the key.

SELECT - INTO AN INTERNAL TABLE

```
SELECT
  [SINGLE [FOR UPDATE]]
  {f1 f2 f3,..| * } [PACKAGE SIZE n1]
  FROM [dbtab | (dbtabname)]
  {APPENDING | INTO}
  [CORRESPONDING FIELDS OF] TABLE [ itab | (itabname) ]
  [UP TO n2 ROWS]
  [BYPASSING BUFFER]
  [WHERE...]
  [ORDER BY...].
```

Creates a record in `itab` for each record in `dbtab` that matches the **WHERE** condition. If you use **PACKAGE SIZE** then `n1` records at a time will be transferred to `itab` during each **SELECT** pass, overwriting the current records or **APPENDING** to the existing ones. Moves selected fields left-to-right into fields of `itab`, or into **CORRESPONDING FIELDS OF** `itab`. `itab` is the literal name of the internal table, and `itabname` is a variable whose value is the name of the internal table. Notice - no **ENDSELECT** is needed. The **SINGLE** statement selects the first matching record. The order of the **FROM** and **APPENDING** or **INTO** clauses is immaterial. **BYPASSING BUFFER** is unnecessary with **SINGLE**.

SELECT - AGGREGATES

```
DATA: w_cnt TYPE I, w_cntd TYPE I,
      w_cntall TYPE I, w_avg LIKE dbtab-f1...
SELECT
  AVG(f1)
  COUNT(f1)
  COUNT(DISTINCT f1)
  COUNT(*)
  MAX(f1)
  MIN(f1)
  SUM(f1)
  FROM [dbtab | (dbtabname)]
  INTO (w_avg, w_cnt, w_cntd, w_cntall, w_max,
  w_min, w_sum)
  [UP TO n ROWS]
  [WHERE...].
```

Notice - no **ENDSELECT** is needed. Extracts specified aggregate information (that is, sum, average, etc.) of all the table records satisfying the **WHERE** condition (bypassing the buffers) and assigns them to the variables in positional order.

SELECT - GROUPED AGGREGATES

```
DATA: w_cnt TYPE I, w_cntd TYPE I,
w_avg LIKE dbtab-f1...
SELECT f1 [f2...]
  AVG([DISTINCT] f1)
  COUNT(*)
  COUNT(DISTINCT f2)
  MAX(f1)
  MIN(f1)
  SUM([DISTINCT] f1)
  FROM [dbtab | (dbtabname)]
  INTO (w_avg, w_cnt, w_cntd, w_max, w_min, w_sum)
  [UP TO n ROWS]
  [WHERE...]
  GROUP BY f1 [f2...]
  [HAVING {AVE|COUNT|MAX|MIN|SUM}(f1) op f3].
  ...
ENDSELECT.
```

Extracts specified aggregate information (that is, sum, average, etc.) for each combination of the values of the fields listed in the **GROUP BY** clause, of all the table records satisfying the **WHERE** condition (bypassing the buffers). **GROUP BY** fields must be specified in the **SELECT** list. **HAVING** limits the number of aggregate lines generated to those that satisfy the condition specified, for example those whose **SUM**(f1) < f3.

SELECT-OPTIONS

Syntax

```
SELECT-OPTIONS rtab FOR { f1 | (f2) }
[DEFAULT g1 [TO g2] [OPTION op1] [SIGN s1]]
[LOWERCASE]
[MATCHCODE OBJECT mco1]
[MEMORY ID pid1]
[MODIF ID mid1]
[NO-DISPLAY]
[NO-EXTENSION]
[NO INTERVALS]
[OBLIGATORY].
```

Description

Adds “High” and “Low” selection criteria to the pre-existing selection screen using the attributes of field `f1` or of the field whose name is stored in `f2`. **DEFAULT** is the default “Low” value. **TO** is the default “High” value. **OPTION** and **SIGN** are the default “Option” and “Sign” values; see the table below for their range of values. **LOWERCASE** allows lower and upper case.

MATCHCODE OBJECT attaches the matchcode object to the Low field. **MEMORY ID** places the parameter ID value in the Low field. **MODIF ID** assigns screen mod group id mid1 to field f1 (see SCREEN-GROUP1 in LOOP AT SCREEN for information). **NO-DISPLAY** hides the selection from the operator. **NO-EXTENSION** blocks the 'Multiple Selection' option, allowing single entry only. **NO INTERVALS** hides the rtab-High field, allowing single values only (that is, no ranges). **OBLIGATORY** makes the low entry mandatory.

This command creates internal table rtab for field f1 with the structure:

Field	Range or Description
Sign	'I'nclude (default) or 'E'xclude.
Option	EQ (default), NE, CP, NP, GE, LT, LE, GT, BT, NB
Low	Comparison value, and inclusive low BT | NB value. This field will have the same type as f1.
High	Inclusive high BT | NB value. This field will have the same type as f1.

See also *Operators*, AT SELECTION-SCREEN, PARAMETERS, RANGES, SELECTION-SCREEN

SELECTION-SCREEN

Syntax

```
SELECTION-SCREEN
{{BEGIN | END} OF LINE
|BEGIN OF BLOCK b1 [WITH FRAME [TITLE text1]]
   [NO INTERVALS]
|BEGIN OF TABBED BLOCK b2 FOR n1 LINES
|TAB (w1) tab1 USER-COMMAND ucomm1
   [DEFAULT [PROGRAM rpt1] SCREEN s1]
|END OF BLOCK b1
|BEGIN OF SCREEN s2 [TITLE text2] [AS WINDOW]
|BEGIN OF SCREEN s3 AS SUBSCREEN [NESTING LEVEL n1]
   [NO INTERVALS]
|END OF SCREEN sn
|COMMENT [/] p(w) text1 [MODIF ID mid1]
|FUNCTION KEY k
|POSITION [p| POS_LOW | POS_HIGH] [MODIF ID mid1]
|PUSHBUTTON [/] [p] text1 USER-COMMAND cmd1
   [MODIF ID mid1]
|SKIP n
|ULINE [/] [p| POS_LOW | POS_HIGH] [(w)]
   [MODIF ID mid1]}.
```

Description
Commands for creating a complex selection screen which the report (program) presents to the user when it begins. Blocks may be nested up to 5 deep. **WITH FRAME** draws a frame around the block on the screen. **TITLE** places the title `text1` in the center of the top line of the frame. `text1` takes the form {`t1` | `TEXT-nnn`} where `t1` is a text variable assigned in the `INITIALIZATION` event and not declared in `DATA`, and `TEXT-nnn` is the normal language-specific text element from the text pool. **NO INTERVALS** hides `rtab-High` on all **SELECT-OPTIONS** in the block or subscreen; this attribute is inherited by nested blocks **WITH FRAME** and is not inherited otherwise. **TABBED BLOCK** creates a subscreen `n1` lines high on which you can place a tabstrip control (**see** `CONTROLS`). Use **TAB** to insert a page in the tabbed block having the tab name `tab1` of width `w1` and assigned the command `ucomm1`. **DEFAULT SCREEN** specifies which subscreen displays when this tab is selected. **PROGRAM** specifies a subscreen in another program. Screen `sn` must be numeric and may be up to four digits. **AS WINDOW** defines this screen as a modal dialog box. Use **NESTING LEVEL** to place this subscreen in a framed tabstrip. **/** forces a newline; it is not allowed between **BEGIN OF LINE** and **END OF LINE**. `p` and `(w)` are position and width measured in columns; they are required in the **COMMENT** option. `n` is the number of lines to skip, between 1 and 9; blank is equivalent to 1. **FUNCTION KEY** places up to five "function key" pushbuttons in the toolbar at the top of the screen. You must assign their labels in the system table `SSCRFIELDS-FUNCTXT_01` through `-FUNCTXT_05`. When the operator selects one of these function keys, `SY-UCOMM` is set to `'FC01'..'FC05'` as appropriate. **POSITION** must be between **BEGIN OF LINE** and **END OF LINE**. **POS_LOW** and **POS_HIGH** are the positions of the Low & High fields created by **SELECT-OPTIONS**.
`cmd1` is the command code assigned to `SY-UCOMM` when the operator selects the pushbutton. **ULINE** may be used between **BEGIN OF LINE** and **END OF LINE**. **MODIF ID** `mid1` assigns the screen modification group id `mid1` to the element. **See** `SCREEN-GROUP1` in `LOOP AT SCREEN`. `PARAMETERS` and `COMMENT` statements may be placed on the **SELECTION-SCREEN** line.
See also *ok-codes*, `PARAMETERS`, `SELECT-OPTIONS`

SET BIT

Syntax
```
SET BIT n1 OF f1 [TO g1].
```
Description
Sets the bit in the (one-based) bit position `n1` of field `f1` to 1 or to the value of `g1`. `g1` must be equal 0 or 1 and `n1` must be greater than zero.

SY-SUBRC 0 success
>0 n1 is greater than the length of f1

See also GET BIT

SET BLANK LINES

Syntax

```
SET BLANK LINES {ON | OFF}.
```

Description

Allows or suppresses WRITEing of blank lines; OFF is the default.

SET COUNTRY c1

Syntax

```
SET COUNTRY c1.
```

Description

Sets the representation of decimal points and dates in WRITE commands according to the specifications in table T005X. If c1=SPACE then country is set to that in the current user's master record.

SY-SUBRC 0 c1 was found in T005X or c1=SPACE
4 otherwise

SET CURSOR

Syntax

```
SET CURSOR
{ FIELD f1 [OFFSET c1] [LINE r1]
| LINE r2 [OFFSET c2]
| c1 r3 }.
```

Description

FIELD positions the cursor in the global field whose name is contained in field f1. **OFFSET** c1 is the zero-based offset in columns from the start of the global field. **LINE** r1 is the line number in which the cursor is placed in step-loop (line number) and list processing (SY-LILLI).

LINE r2 places the cursor in step loop line or list line (SY-LILLI). **OFFSET** c2 is the zero-based offset in columns from the start of the line.

c1 r3 places the cursor in the current screen in the specified column and row.

See also GET CURSOR

SET HANDLER

Description

Registers ABAP Objects event handlers..

This book doesn't discuss ABAP Objects programming.
See also *Object-oriented programming*

SET LANGUAGE

Syntax
`SET LANGUAGE lng1.`

Description
Sets the language for all text-pool and other language-specific elements, and applies only to the current program. The language defaults first to the user's login language (SY-LANGU), and if the text is not defined in that language then to the development language of the object.

SET LEFT SCROLL-BOUNDARY

Syntax
`SET LEFT SCROLL-BOUNDARY [COLUMN c1].`

Description
Sets the left edge of the horizontally-scrollable portion of the screen to the current column. `COLUMN c1` sets it to column `c1`. Applies only to the current screen (Insert this statement in the `TOP-OF-PAGE` event). You can block scrolling of subsequent lines by using `NEW-LINE NO-SCROLLING` (such as for titles and indented comments).
See also `TOP-OF-PAGE`, `NEW-LINE`

SET LOCALE LANGUAGE

Syntax
`SET LOCALE LANGUAGE lg [COUNTRY c] [MODIFIER m].`

Description
Sets the text environment by assigning values to `lg`, and optionally to region `c` and modifier `m`. The fields must be type C with lengths and values found in table TPC0C. The text environment determines how text strings are sorted in itabs and datasets.
See also `CONVERT` (*text*), `GET LOCALE LANGUAGE`

SET MARGIN

Syntax
`SET MARGIN c1 r1.`

Description
Sets the left margin to `c1` and the top margin to `r1` for lists in the current report.

SET PARAMETER ID

Syntax
`SET PARAMETER ID key FIELD f1.`

Description
Assigns the value of f1 to the PID key in the user's SPA/GPA Memory Area.
See also GET PARAMETER ID, *SPA/GPA Memory Area*

SET PF-STATUS

Syntax
```
SET PF-STATUS statusID
[EXCLUDING fnc1 | itab] [IMMEDIATELY].
```
Description
Sets the current GUI status (interactive screen) to statusID. **EXCLUDING** fnc1 deactivates the function fnc1 defined in that status. **EXCLUDING** itab deactivates the functions listed in itab, one per record. **IMMEDIATELY** in list processing forces the new status to take effect at the current list level (in screen processing the new status is set immediately without this option).
SY-PFKEY contains the identity of the current GUI status
See also *pf-status*

SET PROPERTY

Syntax
```
SET PROPERTY OF obj1 p1 = f1 [NO FLUSH].
```
Description
Sets attribute p1 of the OLE2 object obj1 to the value of field f1. **NO FLUSH** continues OLE2 bundling even if the next statement isn't an OLE2 command.

SY-SUBRC 0 all OLE2 commands in the bundle were successful
1 communication error, described in SY-MSGLI
2 method call error, described in dialog box
3 property set up error, described in dialog box
4 property read error, described in dialog box

See also CALL METHOD, CREATE OBJECT, FREE OBJECT, GET PROPERTY

SET SCREEN

Syntax
```
SET SCREEN scr1.
```
Description
Sets the next screen number in an online dialog. Use scr1 = 0 to return to the calling CALL SCREEN.

SET TITLEBAR

Syntax

SET TITLEBAR t1 **[WITH** v1 [v2...]] .

Description

Sets the current title bar at the top of the GUI status to t1. The contents of t1 may be up to 70 characters wide. Create and edit title bar text in the Menu Painter transaction /SE41 **Title List.* Up to nine ampersands (&) in the title text will be replaced by parameters v1, v2....To show a literal ampersand in the title enter two of them, that is, &&.

SY-TITLE contains the current title bar text

SET USER-COMMAND

Syntax

SET USER-COMMAND f1.

Description

In report list generation when the list is to be displayed, this command executes the system function code f1 or AT USER-COMMAND f1 or AT LINE-SELECTION, with the value of f1 assigned to SY-UCOMM, depending on the contents of f1. The system behaves exactly as if the operator had typed f1 into the command field and pressed <Enter> at the current list and cursor positions.

See *okcodes.*

SHIFT

Syntax

SHIFT f1 **[RIGHT] [CIRCULAR] [BY** n1 **PLACES | UP TO** f2**].**

Description

Shifts the string in f1 to the left or right (RIGHT) by the amount in n1 for n1 > 0. If n1 is greater than the width of f1, then f1 is filled with blanks, except for **CIRCULAR**. **CIRCULAR** rotates to the declared width, not the string width.

Example

```
DATA: s1(8) VALUE 'ABCDE_ _ _',   ("ABCDE" & 3 blanks)
      s2(8), s3(8), s4(8).
s2 = s3 = s4 = s1.
SHIFT s1.                          → s1 = 'BCDE_ _ _ _'
SHIFT s2 BY 3 PLACES CIRCULAR.     → s2 = 'DE_ _ _ABC'
SHIFT s3 UP TO 'CD' CIRCULAR.      → s3 = 'CDE_ _ _AB'
SHIFT s4 RIGHT.                    → s4 = '_ABCDE_ _'
```

SY-SUBRC 0 f2 was found (**UP TO** only)
4 otherwise

See also the other string processing commands: CONCATENATE, CONDENSE, OVERLAY, REPLACE, SEARCH, SPLIT, STRLEN(), TRANSLATE

SHIFT ... DELETING

Syntax

SHIFT f1 **{LEFT DELETING LEADING**
| RIGHT DELETING TRAILING} f2.

Description

Shifts the string in f1 left or right until the first or last character in f1 is any character in f2. If no character in f2 appears in f1, then no shift takes place.

Example

```
DATA: s1(8) VALUE 'ABCDEFGH',
      s2(8) VALUE 'ABCDEFGH',
      s3(2) VALUE 'CF'.
SHIFT s1 RIGHT DELETING TRAILING s3. → s1 = '_ _ABCDEF'
SHIFT s2 LEFT DELETING LEADING s3.   → s2 = 'CDEFG_ _'
```

See also the other string processing commands: CONCATENATE, CONDENSE, OVERLAY, REPLACE, SEARCH, SPLIT, STRLEN(), TRANSLATE

SIGN

Syntax

SIGN(x).

Description

Returns the sign of any number x.

0 returns 0

> 0 returns 1

< 0 returns -1

See also *Arithmetic functions*

SIN

Syntax

SIN(y).

Description

Returns the sine of floating point number y, for y in radians.

See also *Arithmetic functions*

SINH

Syntax

SINH(y).

Description

Returns the hyperbolic sine of floating point number y.

See also *Arithmetic functions*

SKIP

Syntax

SKIP [n1 | **TO LINE** r1].

Description

Skips 1 line or n1 lines in a report. **SKIP** is ignored at the beginning of a page except for the first page of a list and a **NEW-PAGE** page. If n1 plus the current line number is greater than the **LINE-COUNT** value less the reserved number of lines, then a new page is started. **TO LINE** r1 jumps up or down to the (one-based) line number r1 of the report; if r1 is greater than the **LINE-COUNT** value there's no **SKIP**.

See also ULINE, WRITE

Sleep

Use the Function Module RZL_SLEEP to pause the program a defined number of seconds. Hint: to insert function module calls into your program, use [Pattern.

SORT

Syntax

`SORT [DESCENDING] [BY f1 [f2...] | fg1].`

Description

Sorts the extracted dataset ascending (or **DESCENDING**) by all the fields in the `HEADER` field-group. **BY** `f1...` sorts by the listed fields in the `HEADER` field-group. **BY** `fg1` sorts by all the fields that are in both the `HEADER` field-group and in field-group `fg1`.

See also `EXTRACT`, `FIELD-GROUPS`, `INSERT`, `LOOP`

SORT ITAB

Syntax

`SORT itab [DESCENDING | BY f1 [DESCENDING] [f2...]].`

Description

Sorts `itab` ascending (or **DESCENDING**) by all non-`F`, `I`, `P` fields. **BY** `f1...` sorts `itab` by the listed fields (all types are allowed) in ascending order, except for those fields for which **DESCENDING** is specified.

See also `APPEND...SORTED BY`

Sound

There's no `BELL`, `BEEP` or `SOUND` command documented for ABAP.

SPA/GPA Memory Area

Description

the Set Parameter/Get Parameter area in memory set aside to store the user's PIDs.

See `GET PARAMETER`, `SET PARAMETER`

SPACE

Description

ABAP reserved constant = `' '`, usable in these constructions.

`X {EQ | NE | <>} SPACE` (relation)

`X = SPACE.` and `MOVE SPACE TO X.` (assignment)

SPLIT

Syntax

SPLIT s1 **AT** s2 **INTO** {tgt1 tgt2 [tgt3...] | **TABLE** itab}.

Description

Splits string s1 at each location of the delimiter s2 and assigns the substrings in order to the targets tgt1 tgt2... or to successive records of itab. If there are more substrings than targets, the remainder are all assigned to the last target. If a substring is wider than its target, it is truncated into the target. The full width of the DATA declaration of the delimiter is used, regardless of its value. Operands are treated as type C without conversion. Trailing spaces are retained in target assignments, and truncated in itab assignments.

Example

```
DATA: delim(2) VALUE ',', t1(6), t2(6), t3(6), t4(20).
SPLIT 'New Jersey, Texas, California, Ohio, Iowa'
   AT delim
   INTO t1, t2, t3, t4.
```

→ t1 contains 'New Je'

→ t2 contains 'Texas'

→ t3 contains 'Califo'

→ t4 contains 'Ohio, Iowa'

Notice that the effective delimiter is ', ' (comma space) because its declared length is 2.

SY-SUBRC 0 all targets were wide enough
4 if any assignment was truncated

See also the other string processing commands: CONCATENATE, CONDENSE, OVERLAY, REPLACE, SEARCH, SHIFT, STRLEN(), TRANSLATE

SQRT

Syntax

SQRT(y).

Description

Returns the square root of floating point number y, for $y \geq 0$.

See also *Arithmetic functions*

START-OF-SELECTION

Syntax

```
START-OF-SELECTION.
```

Description

Event triggered after the selection screen has been processed and before the logical database (LDB) reader triggers the first GET event. Any code block between the REPORT statement and the first event statement is processed as an implicit **START-OF-SELECTION** event before the explicit event is processed. If the program contains more than one **START-OF-SELECTION** event, the code following each will be processed in the order in which the events appear.

See also END-OF-SELECTION, *Events*, INITIALIZATION

STATICS

Syntax

```
STATICS f1.
```

Description

Declares the static variable in a subroutine (FORM or FUNCTION). The syntax is identical to DATA. A static variable retains its value between subsequent calls to the subroutine.

See also CONSTANTS, DATA, LOCAL, TABLES, TYPES

Status

See *Pf-status*

STOP

Syntax

```
STOP.
```

Description

Cancels all further data selection in GET blocks. No further tables are read. Jumps immediately to the END-OF-SELECTION event.

See also CHECK, CONTINUE, EXIT, LEAVE, REJECT

String comparisons

See *Operators (String Comparison)*

String handling

Description

`'Delimit strings with single-quotes.'`

Use two single quotes for an apostrophe:

`'I said "I can''t go."'` → `I said "I can't go".`

Shifting & substrings:

`s1 = s2+c2(w2).`	Results in `s1` containing the substring of `s2` from zero-based offset `c2` for width `w2`.
`s1+c1(w1) = s2+c2(w2).`	Results in `s1` containing that substring of `s2` at zero-based offset `c1` for width `w1`.

Example

```
s1 = '1234567890'.
S2 = 'ABCDEFGHIJ'.
s1+5(3) = s2+2(1).
```

→ `s1` contains `'12345C 90'`

String processing commands

`CONCATENATE`, `CONDENSE`, `OVERLAY`, `REPLACE`, `SEARCH`, `SHIFT`, `SPLIT`, `STRLEN()`, `TRANSLATE`

STRLEN

Syntax

```
STRLEN( s1 ).
```

Description

Returns the width of `s1` to the last non-space character. The argument `s1` must be separated from the parentheses by spaces.

Example

```
DATA w(3) TYPE I, name(36) VALUE 'ABAP Computer Book'.
w = STRLEN(name).
```

→ `w` contains `18`

See also the other string processing commands: `CONCATENATE`, `CONDENSE`, `OVERLAY`, `REPLACE`, `SEARCH`, `SHIFT`, `SPLIT`, `TRANSLATE`

SUBMIT

Syntax

```
SUBMIT [ rpt1 | (rptvar) ]
[AND RETURN [EXPORTING LIST TO MEMORY | TO SAP-SPOOL]]
[LINE-SIZE w1]
[LINE-COUNT n1]
[USING SELECTION SET v1]
[USING SELECTION-SETS OF PROGRAM rpt2]
[VIA SELECTION-SCREEN]
[WITH { p1 OP f1 [SIGN s1]
      | p1 [NOT] BETWEEN f2 AND f3 [SIGN s2]
      | p1 IN rtab}].
```

Description

Runs ABAP report `rpt1` or the report whose name is contained in the variable `rptvar` without returning to the calling program. **`AND RETURN`** returns to the calling program upon completion. **`USING SELECTION SET`** specifies a variant. **`USING SELECTION-SETS`** refers to the `PARAMETERS` and `SELECT-OPTIONS` of `rpt2`; they must be identical to those for the called program `rpt1`. Typically `rpt2` is the calling program; you can use `SY-REPID` for `rpt2`. **`VIA SELECTION-SCREEN`** re-displays the selection screen of the calling report, permitting the user to edit the original entries. `OP` is any operator defined in Operators (`EQ` | `NE` | `GT` | `GE` | `LT` | `LE` | `CP` | `NP`). `s1` and `s2` = `'I'` (include - the default) or `'E'` (exclude). `rtab` is an `itab` which you may define with the `RANGES` command. `LEAVE` in the called report will return immediately to the calling program.

See also *Execute a program*, `CALL DIALOG`, `CALL FUNCTION`, `CALL TRANSACTION`

Subroutine

See `FORM`, `FUNCTION`

SUBTRACT

Syntax

```
SUBTRACT a FROM b.
```

Description

Equivalent to: `b = b - a.` Non-numeric fields are converted. See "Type Conversions" for conversion information.

See also `ADD`, `DIVIDE`, `MOVE`, `MULTIPLY`

SUBTRACT-CORRESPONDING

Syntax

SUBTRACT-CORRESPONDING array1 **FROM** array2.

Description

If array1 & array2 are structured work areas such as header lines, then this command subtracts like-named fields in array1 from array2. Equvalent to

SUBTRACT array1-f1 **FROM** array2-f1.
SUBTRACT array1-f2 **FROM** array2-f2.
etc.

See also ADD-CORRESPONDING, DIVIDE-CORRESPONDING, MOVE-CORRESPONDING, MULTIPLY-CORRESPONDING

SUM

Syntax

SUM.

Description

In a LOOP AT itab ...ENDLOOP this command sums all the numeric (F, I, P) fields in the itab over the range defined by the condition block in which the **SUM** command is found, and places the sums in the header line of itab or the explicitly declared work area wa of the itab. Specifically, in AT FIRST and AT LAST blocks it produces grand totals of the fields, and in AT END OF f1 and AT NEW f1 blocks it produces subtotals of the fields for all records having the current value of f1. For extracted datasets, use SUM().

See also AT FIRST, AT END OF, AT NEW, AT LAST, LOOP AT itab

SUM()

Syntax

SUM(keyfield).

Description

A system function available in the LOOP structure on a sorted extracted dataset where keyfield is a numeric field. Within the condition block AT LAST...ENDAT, **SUM**(keyfield) returns the sum of the values of keyfield in the extract, where keyfield is numeric (type F I P). Within the condition block AT END OF testfield...ENDAT, **SUM**(keyfield) returns the sum of the values of keyfield for the current value of testfield in the extract. For LOOP AT itab see SUM.

See also AT END OF, AT LAST, CNT(), EXTRACT, FIELD-GROUPS, INSERT, LOOP, SORT

SUMMARY

Syntax

`SUMMARY.`

Description

Sets the output format to the "summary" mode. Use instead `FORMAT INTENSIFIED ON` for clarity.

SUPPLY

Syntax

`SUPPLY k1 = f1 k2 = f2 ... TO CONTEXT c1.`

Description

Populates the key fields of the context with the values of `f1 f2`... so the context then determines the dependent values that you can subsequently extract with a `DEMAND` statement.

See also `CONTEXTS`, `DEMAND`

SUPPRESS DIALOG

Syntax

`SUPPRESS DIALOG.`

Description

In a PBO, **`SUPPRESS DIALOG`** stops showing the current screen. Dialog continues as usual, and the next screen will be shown.

Switch command

See `CASE`

System

See *SAPSYSTEM*

System fields

SAP systems maintain a runtime array called `SY` that contains a lot of data about the system and the program that's running. You can get the values of these fields (and assign values to some of them) by referring to them as `SY-XXXXX`. For a list of the system fields see *Appendix A*.

Table Types

Name	Type	Description
Transparent	dbtab	Database table containing master or transaction data.
Pool	dbtab	Obsolete in 3.1 and beyond, bulk storage of many logical tables in a database table.
Internal	itab	Runtime table defined in the current program.
View	dbtab	A hierarchy of transparent tables, frequently used to link transaction data and master data.
Cluster	dbtab	Obsolete in 3.0 and beyond, flat table with "hierarchical" appearance.
Structure	(dbtab)	A table structure with no contents, such as `BDCDATA`. Used to create itabs.

TABLES

Syntax

```
TABLES: dbtab, *dbtab.
```

Description

Establishes a link to the named database table, view or structure (created in /SE11), and declares a header line with the same name and structure as the linked object. `*dbtab` is a second table-cursor pointing to `dbtab`, establishing it as an independent work area.

See also `CONSTANTS`, `DATA`, `LOCAL`, `STATICS`, `TYPES`, *Work Areas*

TAN

Syntax

```
TAN( y ).
```

Description

Returns the tangent of floating point number `y`, for `y` in radians.

See also *Arithmetic functions*

TANH

Syntax

```
TANH( y ).
```

Description

Returns the hyperbolic tangent of floating point number `y`.

See also *Arithmetic functions*

TemSe

Description

"Temporary Sequential" file that contains the actual data for the spooler to send to the printers.

Templates

See *Patterns*

TEMU

Description

(SAPTEMU) obsolete term for SAPGUI.

Text Elements

Description

Each program has an associated pool of text elements in which you can store strings in several languages that will be presented to the user. If the user's log-in language is included in the text pool, then the strings will be presented in that language, otherwise in the development language. Maintain the text pool in /SE38 [Text elements or from the ABAP Editor {GoTo {Text Elements. The text pool can contain the following types of strings:

- Report or program titles
- List headings
- Column headings
- Selection texts (screen labels for `PARAMETERS` and `SELECT-OPTIONS`)
- Numbered Text (constant text passages)

Insert a numbered text element in code, screens, etc. in either of two ways where nnn is its number:

- `TEXT-nnn`
- `'descriptive string'(nnn)`
- (The descriptive string will be inserted if `nnn` is empty.)

See also `DELETE TEXTPOOL`, `INSERT TEXTPOOL`, `READ TEXTPOOL`

Text environment

See `CONVERT` (*text*), `GET LOCALE LANGUAGE`, `SET LOCALE LANGUAGE`

"Thermometer" - progress indicator

See `MOVE ... PERCENTAGE`

Time

Time is stored as a packed field containing the number of seconds since midnight (or before midnight for negative values). Time arithmetic adds and subtracts seconds modulo 86400 (the number of seconds in a day).

See `GET TIME`, `GET RUN TIME`, `SY-UZEIT`, `SY-TZONE`, `SY-DAYST`

Timestamp

See CONVERT *(timestamp)*, GET TIME STAMP, WRITE...TIME ZONE

Time zone

See CONVERT *(timestamp)*, GET TIME STAMP, WRITE...TIME ZONE

Titlebar

See SET TITLEBAR

TOP-OF-PAGE

Syntax

TOP-OF-PAGE [DURING LINE-SELECTION].

Description

These events are triggered in interactive reports at the beginning of each internally-generated page break, after the standard heading is written and before the first data are processed. **TOP-OF-PAGE** is triggered in the basic list. **TOP-OF-PAGE DURING LINE-SELECTION** is triggered in lists generated interactively (detail lists). These events are not triggered by NEW-PAGE.

See also END-OF-PAGE, *Events*

Transaction Codes

See *Appendix B*

TRANSFER

Syntax

TRANSFER array1 **TO** filename **[LENGTH** w1**].**

Description

Transfers the contents of array1 to the sequential file filename on the application server. If filename is not open, it attempts to OPEN it FOR OUTPUT IN BINARY MODE. array1 may be a field string or a table work area. **LENGTH** returns the length of the transferred record in w1. A newline character is appended to the end of each record if the file was opened IN TEXT MODE. Note: Trailing blanks are truncated when writing in text mode. To create fixed-length records use binary mode to write each record and append your own line break (CRLF - Hex 0D0A).

See also OPEN DATASET, READ DATASET, CLOSE DATASET, DELETE DATASET

TRANSLATE

Syntax

```
TRANSLATE string1
{ TO {UPPER | LOWER} CASE
| USING string2
| FROM CODE PAGE cp1 TO CODE PAGE cp2
| FROM NUMBER FORMAT nf1 TO NUMBER FORMAT nf2}.
```

Description

Alters `string1` in any of several ways:

`TO {UPPER | LOWER} CASE` forces all characters to upper or lower case.

`USING string2` replaces in `string1` every case-sensitive example of the first character of each pair in `string2` with the second character in the pair.

Example

```
DATA date1(10) VALUE '12-31-2001.
TRANSLATE date1 USING '-/./'. → date1 contains
'12/31/2001'
```

`{ FROM | TO } CODE PAGE` converts between SAP character codes and those in the named code page. Types `I`, `P`, `F`, `X` are not converted. Use `/SPAD` to maintain the conversion tables `TCP00-TCP02`.

Example

```
TRANSLATE f1 FROM CODE PAGE '1110' TO CODE PAGE '0100'.
```

converts the contents of `string1` from the HP character set to EBCDIC;

`{ FROM | TO } NUMBER FORMAT` Converts between SAP Type I & F number formats and those of HP, IBM, SINEX (nf= '0000') or DEC Alpha OSF (nf= '0101').

See also the other string processing commands: `CONCATENATE`, `CONDENSE`, `OVERLAY`, `REPLACE`, `SEARCH`, `SHIFT`, `SPLIT`, `STRLEN()`

True

Description

There's no logical type in ABAP. Logical state is frequently represented by a one-character TYPE C field with its initial value `' '` or `SPACE` for `FALSE`, and `'X'` for `TRUE`.

TRUNC

Syntax

```
TRUNC( x ).
```

Description

Returns the truncated value of `x`, that is, the integer portion of any number `x`.

See also *Arithmetic functions*

Type

Description

The primitive data types in ABAP are:

Type Symbol	Meaning	Initial Value	Default Bytes	Allowed Bytes	Value Range
C	Text (default)	SPACE	1	②	
D	Date YYYYMMDD	'00000000'	8	8*	≥ 01/01/1900
F	Float	'0.0'④	8①	8①*	±1E-307 to ±1E+308
I	Integer	0000	4①	8①*	-2**31 to +2**31-1
N	Numeric text	'0'	1	②	②
P	Packed	00000000	8	1-16	③
T	Time HHMMSS	'000000'	6	6*	-86,399 to +86,399 sec.
X	Hex	00	1	②	⑤
STRING	Text string	Empty string	NA	Any number	
XSTRING	Byte String	Empty string	NA	Any number	
CURSOR	Table Cursor			*	

Value range is for numeric fields only.
* = no width specification required
① Machine-specific; typical width shown.
② 1 to approximately 64k bytes.
③ Packed resolution = 2 x length - 1, up to 14 decimal places.
④ Enclose float numbers in quotes or SAP will interpret the decimal point as a command terminator. Float resolution is 15 decimal places. They may be entered in direct or scientific notation: '-1234.56' or '+1.23456E3' or '4' or '52E-12' etc.
⑤ Hex resolution = 2 x length.

There's no logical type in ABAP. Logical state is frequently represented by a one-character TYPE C field with its initial ' ' value or `SPACE` for `FALSE`, and `'X'` for `TRUE`.

Use packed numbers for monetary fields, and use integer fields for indexes, column numbers, positions, etc.

If you're doing hexadecimal operations that will change the sign of any TYPE X variable, then make sure that variable is exactly four bytes wide. The sign bit may not be properly handled if the field is wider or narrower.

Example

```
DATA: x4(4) TYPE X, x6(6) TYPE X.
x4 = '00006666'.
x6 = - x4.
WRITE: / x4, x6.  →00006666      0000FFFF999A
```

(that is, the leading `F` is not carried forward beyond the fourth byte)

The data dictionary contains several derived data types which can be used in table domain data type definitions, including:

Derived Type	Primitive Equivalent	Description
ACCP	N6	Posting period in the form YYYYMM
CHAR n	Cn	Character string, n ≤ 255
CLNT	C3	Client (Mandt), the first key field in every client-dependent table
CUKY	C5	Set of possible currencies; referenced by CURR
CURR n,m,s	⑥	Currency field; refers to CUKY
DATS	D	Date field YYYYMMDD
DEC n,m,s	⑥	Numeric field with decimal point & perhaps sign & commas
FLTP	F	Floating point field with 8 bytes accuracy
INT1	1-Byte int.	Integer, 0 to 255
INT2	2-Byte int.	Integer, 0 to 65,536
INT4	I	Integer, -2**31 to +2**31-1
LANG	C1	Language (1-byte internal, 2 bytes external representation)

Derived Type	Primitive Equivalent	Description
`LCHR n`	`Cn`	Character string, 256 to 65,356 bytes long; must be the last field in a table and be preceded by a field of type `INT2` to carry its length; may not be used in a `WHERE` condition.
`LRAW n`	`Xn`	Uninterpreted (raw) string, 256 to 65,356 bytes long; must be the last field in a table and be preceded by a field of type `INT2` to carry its length; may not be used in a `WHERE` condition.
`NUMC n`	`Nn`	Numeric character field; may contain only digits, 1 to approximately 64k bytes long
`PREC`	`X2`	Sets the precision of the `QUAN` field
`QUAN n,m,s`	⑥	Contains quantities; refers to `UNIT`
`RAW n`	`Xn`	Uninterrupted string, n ≤ 255
`RAWSTRING`	`XSTRING`	Uninterrupted byte string
`STRING`	`STRING`	Uninterrupted character string
`TIMS`	`T`	Time, `HHMMSS`
`UNIT n`	`Cn`	Set of possible quantity units of measure; points to table T006

⑥ = `P((n+2)/2) DECIMALS m [NO SIGN]`

Type Conversions

SAP automatically converts field types when it can.
See *Appendix D* for a table showing the conversions

TYPE-POOL

Syntax

```
TYPE-POOL t1.
```

Description

A type-pool is a code block that contains type and constant definitions you can include in your programs. You create it by /SE80 *Dictionary Objects [Edit <Type Group = `'typepoolname'`.

Example

```
TYPE-POOL t1.
TYPES: f1(8), f2(4) TYPE I, f3(8) TYPE P DECIMALS 2.
CONSTANTS: newline TYPE X VALUE '0D'.
```

See also CONSTANTS, TYPE-POOLS, TYPES

TYPE-POOLS

Syntax

```
TYPE-POOLS t1.
```

Description

Includes in the current program the types and constants of type group `t1` that were defined in **TYPE-POOL** `t1`. Similar to an INCLUDE for types and constants. Type groups are maintained in `/SE11`.

See also TYPE-POOL, TYPES

TYPES

Syntax

```
TYPES u1[(w1)] { TYPE t1 | LIKE f1 [OCCURS n1]}
[DECIMALS d1].
```

or

```
TYPES:
  BEGIN OF u2,
    u21...,
    ...,
    u2n...,
  END OF u2.
```

or

```
TYPES u3 {TYPE array1 | LIKE array2} OCCURS n3.
```

or

```
TYPES itabtype
{TYPE kind1 OF arraytype1 | LIKE kind2 OF array1}
[WITH {UNIQUE | [NON-UNIQUE]}
      {KEY k1 k2 kn | KEY TABLE_LINE | DEFAULT KEY}]
[INITIAL SIZE n1].
```

or

Use either of the following techniques to declare a RANGES table type (see RANGES).

```
TYPES rtabtype LIKE RANGE OF f1 [INITIAL SIZE n1].
```

or where `ty1` is any standard type.

```
TYPES rtabtype TYPE RANGE OF ty1 [INITIAL SIZE n1].
```

Description

Creates runtime data types which may be used in any way the standard types may be used. `w1` is the specified width of the field, applicable only to Types `C` ($w \leq 65535$), `N` ($w \leq 65535$), `P` ($w \leq 16$) and `X` ($w \leq 65535$). `t1` may be any standard or previously-defined type. `f1` may be a dbtab field, data dictionary field or system field (see *Appendix A*), or any field you have already defined in a DATA statement. **OCCURS** creates an itabtype with no header line whose structure is type t1 or matches that of f1. **TYPE LINE OF** creates an array whose structure matches the itabtype of itype1. **LIKE LINE OF** creates an array whose structure matches that of itab1. **DECIMALS** applies only for Type P. **BEGIN OF... END OF** creates an array whose elements may have any standard types or any of the user-defined types shown above. Refer to elements of the array using arrayname hyphen elementname, for example u2-u21.

kind1 and kind2 is one of [STANDARD TABLE], SORTED TABLE, HASHED TABLE. The standard table is the default and must be non-unique. The hashed table must be unique, and the sorted table may be either. Fields k1, k2... are the selected keys, searched in the order listed. **TABLE_LINE** is the entire line of the table as a string. **DEFAULT KEY** is the "standard" key definition which consists of all the table's fields which are not numeric (F, I, P) and not internal tables themselves, in the order in which they appear in the line.

The **OCCURS** value must be a literal number; zero is allowed. Its value is effective only in the command APPEND SORTED BY. The later **INITIAL SIZE** parameter specifies the number of lines in the initial itab, and zero is permitted. In both cases, the itab size is dynamically adjusted in operations.

See also CONSTANTS, DATA, LOCAL, STATICS, TABLES, TYPE-POOLS

ULINE

Syntax

ULINE [AT] [/][p][(w)][NO-GAP].

Description

Writes an underline in a report, optionally starting at position p, for width w. "/" starts a new line. **ULINE** without position and length parameters automatically starts a new line. p and w represent position and length literals or variables, measured in columns. The parameters /, p and w must appear in that order with no intervening spaces. **NO-GAP** suppresses the following blank. If you use the **AT** addition, then p and w may be literals or variables, measured in columns. Without that addition, they must be literals.

See also SKIP, WRITE

UNPACK

Syntax

```
UNPACK p1 TO c1.
```

Description

Unpacks the packed field `p1` and places the unpacked value in the character field `c1`. The target is filled with leading zeros if it is longer than `p1`, and is left-truncated if shorter. The sign of `p1` is not placed in `c1`. If `p1` is not a packed type, then it is converted to Type C following the rules described in Type Conversions. A packed field stores two digits per byte in Binary-Coded Decimal (BCD) format, reserving the first nibble (half-byte) for the sign, so its resolution is twice its length minus 1.

UPDATE

Syntax

```
UPDATE dbtab [CLIENT SPECIFIED]
[ FROM wa | FROM TABLE itab
| SET st1 [st2...] [WHERE <condition>] ].
```

Description

Updates values in the database table `dbtab` from the header record, overwriting the record if the key matches, and inserting the record if there's no matching key. The command normally affects only the current client records. If you use **`CLIENT SPECIFIED`** then client (MANDT) becomes a normal field and you can affect other client's records. **`FROM`** updates from the work area `wa`. **`FROM TABLE`** updates all records in `dbtab` whose keys match those in `itab`. **`SET`** updates all `dbtab` records for which `condition` is true by executing the statements st1... of the forms:

```
f1 = g1.
f2 = f2 + g2.
f2 = f2 - g2.
```

Where `f1` is any type field in `dbtab` and `f2` is a numeric field in `dbtab`, and where `g1` is any type variable or literal and `g2` is a numeric variable or literal. If `f2` is `NULL` (hex `00`) then it is not changed (**`NULL`** is not the initial value for most Types).

`SY-SUBRC`	0 successful
	0 `itab` is empty or every record in `itab` was updated
	4 otherwise
`SY-DBCNT`	contains the number of fields updated

See also `INSERT`, `MODIFY`, `WHERE`

Upper case

See `TRANSLATE`

`UPLOAD`

Description

Function module to read a local disk file from the user's workstation into an internal table. Hint: to insert function module calls into your program, use [Pattern. This function module presents the user with a dialog box to enter filename and filetype. See `WS_UPLOAD` for details; it is similar except filename and type are parameters rather than prompts.

See also `DOWNLOAD`, `WS_DOWNLOAD`, `WS_EXECUTE`, `WS_QUERY`, `WS_UPLOAD`

User Exits

Description

Enhancements to the native applications may be installed at the hooks provided in the SAP programs called user exits. You manage some types of user exits in transactions `/SMOD` and `/CMOD`. The stub pools for user exits are typically INCLUDE programs with names ending in "ZZ". See On-line help: [*Basis Components* [*ABAP Development Workbench* [*Modifications and Improvements* for more information.

`VARY`

Syntax

`VARY[ING]` `v1` **`FROM`** `array1-fm` **`NEXT`** `array1-fn.`

Description

Options on **`DO`**...(**`VARYING`**) and **`WHILE`**...(**`VARY`**) that change the variable v1 in subsequent passes. On the first pass `v1` is assigned the value of element `fm` in `array1`. On the second pass `v1` is assigned the value of element `fn`. For each subsequent pass, it is assigned the value of the `array1` element that is further down `array1` by the same separation as that between `fm` and `fn`. The elements must be compatible with and convertible to the type of the variable. If `v1` is changed during the pass, the corresponding element in `array1` will be assigned the new value of `v1`, unless the pass terminates in a dialog message. The option can assign any number of elements. The parent command can have any number of **`VARY`** | **`VARYING`** options. See *Appendix G.6* for an example.

Verbuchen

Description

"Verbuchen" is the German word for "enter" (as entering invoices in the books), "post", and "update". In R/3 it refers to update work processes which are identified as "VB" types. Update components are labeled either as U1 and U2, or as V1 and V2.

See also *Work processes*

wa

See *Work Area*

Wait

Use the Function Module `RZL_SLEEP` to pause the program a defined number of seconds. Hint: to insert function module calls into your program, use `[`Pattern.

WHERE

Syntax

```
WHERE <condition>.
```

Description

Qualifier that may be used in `DELETE`, `OPEN CURSOR`, `SELECT` and `UPDATE` statements. The parent command processes only those records that satisfy `<condition>`.

See *Condition*

WHILE

Syntax

```
WHILE <condition>
[VARY v1 FROM array1-fm NEXT array1-fn …].
  ...
ENDWHILE.
```

Description

Processes the code block repeatedly while the condition is true or until terminated by `EXIT`. `CONTINUE` unconditionally skips to the **`ENDWHILE`** for the next iteration. `CHECK <condition>` skips to **`ENDWHILE`** for the next iteration if the condition is false. **`VARY`** steps the variable v1 in subsequent passes, see `VARY` for an example. `SY-INDEX` contains the one-based loop count for the current nest level. After the **`ENDWHILE`** it is restored to the value it had before the **`WHILE`**.

See also `CASE`, *Condition*, `DO`, `IF`, `LOOP`, `VARY`

Wildcards

See *Patterns*

WINDOW

Syntax

```
WINDOW STARTING AT c1 r1 [ENDING AT c2 r2]
[WITH FRAME [TITLE t1]].
```

Description

In list processing, **WINDOW** places a modal dialog box on the list. Subsequent WRITE statements appear in the dialog box until the end of the current event. **WITH FRAME** surrounds the window with a frame. **TITLE** places title t1 in the center of the top of the frame. Use the Screen Painter selection *List_in_a_Dialog_Box if you are using a GUI status.

See also CALL SCREEN

Work area

A work area is a runtime field array that matches the structure of a database or internal table. You can manually define a work area in a DATA statement using the INCLUDE STRUCTURE tablename or LIKE tablename constructions. Or you can manually define a work area in a DATA statement by explicitly calling out all the fields with their widths and types.

See also *Header Line*

Work processes

Description

Work processes are SAP programs set up to process user and system requests. They include:

Description	Type
Dialog work processes (user activities)	DIA
Background work processes	BTC
Update work processes	VB (see Verbuchen)
Spool work processes	SPO
Enqueue (locking) work processes	ENQ

See also *Dispatcher, Instance, Verbuchen*

WRITE

Syntax

```
WRITE [[AT] [/][p][(w)]] f1 [fmt1] [attr1]
      [AS {CHECKBOX | SYMBOL | ICON | LINE}].
```

Description

Displays field `f1` on report line(s). f1 may be a variable, a table field, a field-symbol, a literal string or a text-element. The "/"and the dimensioning parameters `p` and `w` must appear in the order shown with no intervening spaces. "/" starts a new line. If you use the **AT** addition, then `p` and `w` may be literals or variables, measured in columns. Without that addition, they must be literals. Without dimensioning parameters, the first field starts in column one, and subsequent fields are spaced by one blank. `fmt1` may be one or more of the format options listed below. `attr1` may be any of the attributes discussed in **FORMAT**. **AS CHECKBOX** places an empty checkbox on the screen. The user may click on it to set the field to 'X' which is available in `READ LINE...FIELD VALUE`. **AS SYMBOL** places a symbol on the screen. `INCLUDE` either system include `<SYMBOL>` or `<LIST>` and get the names of the symbols from the include. Most symbols are one character wide. Get the width of any symbol with `DESCRIBE FIELD`. **AS ICON** places an icon on the screen. `INCLUDE` either system include `<ICON>` or `<LIST>` and get the names of the symbols from the include. Most icons are two characters wide. Get the width of any symbol with `DESCRIBE FIELD`. **AS LINE** places a line-draw character on the screen. `INCLUDE` either system include `<LINE>` or `<LIST>` and get the names of the characters from the include. Normally SAP will automatically draw box corners and the like when you write lines with `'_'` or `SY-ULINE` and with `'|'` or `SY-VLINE`. For complex or dense structures, you may need to force corner and intersection characters in some places.

See the table on the next page for descriptions of the format options.

fmt1 Formatting option	**Description**
CENTERED	Show the value centered in its defined width
CURRENCY c1	Show with number of decimals specified in TCURX for currency type c1
DECIMALS d1	For types F I P; rounds up to d1 decimal places or pads with zeros to fill d1 decimal places
DD/MM/YY[YY] | **MM/DD/YY[YY]** | **DDMMYY** | **MMDDYY** | **YYMMDD**	Date field options
EXPONENT e1	For type F; set the exponent to e1, adjust the decimal point of the mantissa to fit
LEFT-JUSTIFIED	Left-justify the value in its defined width (default for Types C D N T X)
NO-GAP	Suppresses the space normally inserted after f1
NO-SIGN	The sign is not shown for Types F I P
NO-ZERO	Leading zeros and zero fields are shown as blanks
QUICKINFO t1	Pop up the information text t1 when the cursor alights on this output field. t1 can be up to 40 characters long and must be Type C, N, D or T.
RIGHT-JUSTIFIED	Right-justify the contents in its defined width (default for Types F I P)
ROUND r1	For type P; moves decimal point left (r1>0) or right then displays with **DECIMALS** value. See the example below
TIME ZONE z1	Displays the timestamp field as date and time in the listed time zone. You can also use date field options to alter the form of the date display.
UNDER g1	Aligns with the starting column of field g1 in the previous **WRITE**
UNIT u1	Displays type P with the number of decimals specified in T006 for unit type u1
USING EDIT MASK	Writes the contents of f1 using the pattern contained in the mask. See the table of mask characters and the example below.
USING NO EDIT MASK	Disables the conversion routine associated with the dictionary domain of f1

Example – WRITE … ROUND

```
DATA: p1 TYPE P VALUE '-765.4321' DECIMALS 3.
WRITE: / p1.              →          765.432-
WRITE: / p1 ROUND 0.      →          765.432-
WRITE: / p1 ROUND 1.      →           76.543-
WRITE: / p1 ROUND 2.      →            7.654-
WRITE: / p1 ROUND 3.      →            0.765-
WRITE: / p1 ROUND 4.      →            0.077-
WRITE: / p1 ROUND -1.     →        7,654.320-
WRITE: / p1 ROUND -2.     →       76,543.200-
WRITE: / p1 ROUND -3.     →      765,432.000-
```

Mask Character	Description
_ (underscore)	Writes one character from `f1`.
`'V'`	Shows the location of the minus sign for Types `P` or `I`.
`'LL'`	At the beginning of the mask left-justifies the output.
`'RR'`	At the beginning of the mask right-justifies the output.
Others	All other characters in the mask appear literally in the output.

Example – WRITE … USING MASK

```
DATA: c1(4) VALUE 'ABCD',
      i1 TYPE I VALUE '-1234',
      p1 TYPE P VALUE '-1234'.
WRITE: /    c1 USING EDIT MASK  '__:__'.     → AB:C
WRITE: /(8) c1 USING EDIT MASK  '__:__'.     → AB:CD
WRITE: /(8) c1 USING EDIT MASK  '__:_'.      → AB:C
WRITE: /(8) c1 USING EDIT MASK  'RR__:__'.   → AB:CD
WRITE: /(8) c1 USING EDIT MASK  'RRR__:__'.  → RAB:CD
WRITE: /(8) c1 USING EDIT MASK  '_a_b_c_d'.  → AaBbCcDd
WRITE: /(8) i1 USING EDIT MASK  '____'.      → 1234
WRITE: /(8) i1 USING EDIT MASK  '____V'.     → 1234-
WRITE: /(8) i1 USING EDIT MASK  'V____'.     → -1234
WRITE: /(8) i1 USING EDIT MASK  'RRV____'.   → -1234
WRITE: /(8) p1 USING EDIT MASK  '____'.          →    1234
WRITE: /(8) p1 USING EDIT MASK  '____V'.         →    1234-
WRITE: /(8) p1 USING EDIT MASK  'V____'.         →   -1234
```

See also POSITION, PRINT-CONTROL, SKIP, ULINE

WRITE...TO

Syntax

WRITE f1 **TO** f3[+p1[(w1)]] [fmt1] [**INDEX** ndx1].

Description

Formats the contents of f1 to specification fmt1, then uses it to overwrite character field f3 or itab work area f3, beginning at offset p1 for width w1. See the fmt1 definitions in WRITE. Parameters p1 and w1 may be literals or variables. If parameter w1 is used, then **WRITE** overwrites just that width of f3. If w1 is not used then the command overwrites the entire width of f3. **INDEX** overwrites record ndx1 of internal table f3.

See also =, MOVE

WS_DOWNLOAD

Syntax

```
CALL FUNCTION 'WS_DOWNLOAD'
  EXPORTING
    BIN_FILESIZE          =''  File length for binary files
    CODEPAGE              =''  'IBM' for Windows targets
    FILENAME              =''  Name of the file
    FILETYPE              =''  File type (see below)
    MODE                  =''  Overwrite ='', Append = 'A'
    WK1_N_FORMAT          =''  Format, spreadsheet value columns
    WK1_N_SIZE            =''  Column width, value columns
    WK1_T_FORMAT          =''  Format, spreadsheet text columns
    WK1_T_SIZE            =''  Column width, text columns
  IMPORTING
    FILELENGTH            =    Quantity of bytes transferred
  TABLES
    DATA_TAB              =    itab containing the data to download
  EXCEPTIONS
    FILE_OPEN_ERROR       = 01   File cannot be opened
    FILE_WRITE_ERROR      = 02   File cannot be written
    INVALID_FILESIZE      = 03   Invalid parameter BIN_FILESIZE
    INVALID_TABLE_WIDTH   = 04   Invalid table structure
    INVALID_TYPE          = 05   Invalid value for parameter
                                 FILETYPE
     NO_BATCH             = 06.  Frontend function cannot be
                                 executed in the background
```

Description

WS_DOWNLOAD is a function module that stores the data from an internal table in a local disk file on the user's workstation. The calling program must provide filename and filetype. DOWNLOAD is a similar function module except it prompts for name and type. Hint: to insert function module calls into your program, use [Pattern.

Filetype options	Description
ASC	Records are terminated by end-of-line characters
BIN	itab must contain at least one TYPE X field and file size must be specified in BIN_FILESIZE
DAT	Excel file format
WK1	Spreadsheet format

See also DOWNLOAD, UPLOAD, WS_EXECUTE, WS_QUERY, WS_UPLOAD

WS_EXECUTE

Syntax

```
CALL FUNCTION 'WS_EXECUTE'
  EXPORTING
    COMMANDLINE          =''  Parameters (command line)
    INFORM               =''  Activating the confirmation from...
    PROGRAM              =''  Path + name of the program
    STAT                 =''  Dialog parameters
    WINID                =''  Dialog parameters
  IMPORTING
    RBUFF                =''  Dialog parameters
  EXCEPTIONS
    NO_BATCH             = 03   Frontend function cannot be
                                executed in the background
    PROG_NOT_FOUND       = 04.  Program couldn't be found to
                                execute
```

Description

WS_EXECUTE is a function module that launches a local program on the user's workstation. The calling program must provide the filename. Hint: to insert function module calls into your program, use [Pattern.

See also DOWNLOAD, UPLOAD, WS_DOWNLOAD, WS_QUERY, WS_UPLOAD

WS_QUERY

Syntax

```
CALL FUNCTION 'WS_QUERY'
  EXPORTING
    ENVIRONMENT          =''  Variable name for environment
                                query
    FILENAME             =''  File name for DE, FE, FL
    QUERY                =''  Query command (see below)
    WINID                =''  Window ID for query WI
  IMPORTING
    RETURN               =''  Result of the query
  EXCEPTIONS
    INV_QUERY            = 01     Incorrect value for QUERY
    NO_BATCH             = 02.    Frontend function cannot be
                                  Executed in the background
```

Description

WS_QUERY is a function module that checks the existence and attributes of a local object on the user's workstation. The calling program must provide filename. Hint: to insert function module calls into your program, use [Pattern.

Query command	Description
CD	Directory
DE	Directory Exists
EN	Environment
FE	File Exists
FL	File Length
GM	GMUX Version
OS	Operating System
WI	Window ID
WS	Window System
XP	eXecute Path

See also DOWNLOAD, UPLOAD, WS_DOWNLOAD, WS_EXECUTE, WS_UPLOAD

WS_UPLOAD

Syntax

```
CALL FUNCTION 'WS_UPLOAD'
  EXPORTING
    CODEPAGE              =''  'IBM' for Windows system
    FILENAME              =''  Name of the file to upload
    FILETYPE              =''  File type (see below)
  IMPORTING
    FILELENGTH            =''  number of bytes transferred
  TABLES
    DATA_TAB              =''  itab receiving the upload
  EXCEPTIONS
    CONVERSION_ERROR      = 01    Error in the data conversion
    FILE_OPEN_ERROR       = 02    File cannot be opened
    FILE_READ_ERROR       = 03    File cannot be read
    INVALID_TABLE_WIDTH   = 04    Invalid table structure
    INVALID_TYPE          = 05    Incorrect FILETYPE
    NO_BATCH              = 06.   Frontend function cannot be
                                  executed in the background
```

Description

WS_UPLOAD is a function module that reads a local disk file from the user's workstation into an internal table. The calling program must provide filename and filetype. UPLOAD is similar except it prompts for file and type. Hint: to insert function module calls into your program, use [Pattern.

Filetype options	**Description**
ASC	Records are terminated by end-of-line characters
BIN	itab must contain at least one TYPE X field and file size must be specified in BIN_FILESIZE
DAT	Excel file format
WK1	Spreadsheet format

See also DOWNLOAD, UPLOAD, WS_DOWNLOAD, WS_EXECUTE, WS_QUERY

York-Mills Notation

See *Appendix I*

Appendix A1 - System Fields (by use)

SAP systems maintain a runtime array called `SY` that contains a lot of data about the system and the program that's running. You can get the values of these fields (and assign values to some of them) by referring to them as `SY-XXXXX` where `XXXXX` is the field name. For example the user's login name is stored in `SY-UNAME`. Releases 2.x and 3.x have the same system fields. Releases 4.0 and 4.x each have slightly different lists of system fields. The fields in this list are valid for all releases through 4.6 unless otherwise identified. 78 of the original system fields are obsolete or for internal use only; those are not shown here.

Field Name	Use	Description	Type	Len
CALLD	ABAP program	Call mode of the ABAP program	CHAR	1
CPROG	ABAP program	Program that called the current external procedure	CHAR	40
DBNAM	ABAP program	Logical database linked to the program	CHAR	20
DYNGR	ABAP program	Screen group of the current screen	CHAR	4
DYNNR	ABAP program	Number of the current screen	CHAR	4
LDBPG	ABAP program	Logical database program for `SY-DBNAM`	CHAR	40
REPID	ABAP program	Current main program	CHAR	40
TCODE	ABAP program	Current transaction code	CHAR	20
BATCH	Background processing	'X' = program is running in the background	CHAR	1
BINPT	Batch input	'X' = program is running in the background	CHAR	1
ABCDE	Constants	Alphabet (A,B,C,...)	CHAR	26
ULINE	Constants	Horizontal line with length 255	CHAR	255
VLINE	Constants	Vertical line (|)	CHAR	1
COLNO	Creating lists	Current list column	INT4	10
LINCT	Creating lists	Page length in a list (from REPORT)	INT4	10
LINNO	Creating lists	Current line	INT4	10
LINSZ	Creating lists	Line width in a list (from REPORT)	INT4	10
PAGNO	Creating lists	Current page	INT4	10
TVAR0	Creating lists	Text variable for ABAP text elements	CHAR	20
TVAR1	Creating lists	Text variable for ABAP text elements	CHAR	20
TVAR2	Creating lists	Text variable for ABAP text elements	CHAR	20
TVAR3	Creating lists	Text variable for ABAP text elements	CHAR	20
TVAR4	Creating lists	Text variable for ABAP text elements	CHAR	20

Field Name	Use	Description	Type	Len
TVAR5	Creating lists	Text variable for ABAP text elements	CHAR	20
TVAR6	Creating lists	Text variable for ABAP text elements	CHAR	20
TVAR7	Creating lists	Text variable for ABAP text elements	CHAR	20
TVAR8	Creating lists	Text variable for ABAP text elements	CHAR	20
TVAR9	Creating lists	Text variable for ABAP text elements	CHAR	20
WTITL	Creating lists	'X' = standard page header	CHAR	1
DBCNT	Database access	Number of database rows processed	INT4	10
DATLO	Date and time	User's local date; releases 4.x only	DATS	8
DATUM	Date and time	Current application server date	DATS	8
DATUT	Date and time	Global UTC (GMT) date; release 4.0 only	DATS	8
DAYST	Date and time	'X' = summer (daylight saving) time	CHAR	1
FDAYW	Date and time	Weekday in the factory calendar	INT1	3
TIMLO	Date and time	User's local time; releases 4.x only	TIMS	6
TIMUT	Date and time	Global UTC (GMT) time; release 4.0 only	TIMS	6
TSTLO	Date and time	Local date and time; release 4.0 only	NUMC	14
TSTUT	Date and time	UTC (GMT) date and time; release 4.0 only	NUMC	14
TZONE	Date and time	Difference between local time and GMT in seconds	INT4	10
UZEIT	Date and time	Current application server time	TIMS	6
ZONLO	Date and time	User's time zone; releases 4.x only	CHAR	6
TABIX	Internal tables	Current line of an internal table	INT4	10
TFILL	Internal tables	Current number of entries in internal table	INT4	10
TLENG	Internal tables	Line width of an internal table	INT4	10
TOCCU	Internal tables	OCCURS parameter with internal tables	INT4	10
STACO	List processing	Number of first displayed column	INT4	10
STARO	List processing	Number of first displayed line on this page	INT4	10
INDEX	Loops	Current loop pass in DO and WHILE loop	INT4	10
MSGID	Messages	Message class(application area)	CHAR	20
MSGLI	Messages	Contents of the message line	CHAR	60
MSGNO	Messages	Message number	NUMC	3
MSGTY	Messages	Message type (A,E,I,S,W,X)	CHAR	1
MSGV1	Messages	Message variable	CHAR	50
MSGV2	Messages	Message variable	CHAR	50
MSGV3	Messages	Message variable	CHAR	50
MSGV4	Messages	Message variable	CHAR	50
CALLR	Printing lists	ID for print dialog function	CHAR	8

Field Name	Use	Description	Type	Len
MACOL	Printing lists	Left columns from the SET MARGIN statement	INT4	10
MAROW	Printing lists	Top rows from the SET MARGIN statement	INT4	10
PAART	Printing lists	Print formatting	CHAR	16
PDEST	Printing lists	Output device	CHAR	4
PEXPI	Printing lists	Spool retention period	NUMC	1
PLIST	Printing lists	Name of spool request	CHAR	12
PRABT	Printing lists	Cover sheet: Department	CHAR	12
PRBIG	Printing lists	'X' = Selection cover sheet	CHAR	1
PRCOP	Printing lists	Number of copies	NUMC	3
PRDSN	Printing lists	Name of the spool dataset	CHAR	6
PRIMM	Printing lists	'X' = Print immediately	CHAR	1
PRNEW	Printing lists	'X' = New spool request (list)	CHAR	1
PRREC	Printing lists	Recipient	CHAR	12
PRREL	Printing lists	'X' = Delete after printing	CHAR	1
PRTXT	Printing lists	Text for cover sheet	CHAR	68
RTITL	Printing lists	Program from which you are printing	CHAR	70
SPONO	Printing lists	Spool number	NUMC	10
CPAGE	Processing lists	Current page number	INT4	10
LILLI	Processing lists	List line selected	INT4	10
LISEL	Processing lists	Contents of the selected line	CHAR	255
LISTI	Processing lists	Index of the selected list (0=base, 1=detail 1 ...)	INT4	10
LSIND	Processing lists	Index of the detail list (0=base, 1=detail 1 ...)	INT4	10
DBSYS	R/3 System	Name of the central database system (e.g. Oracle, Informix, etc.)	CHAR	10
HOST	R/3 System	Name of application server	CHAR	8
LANGU	R/3 System	User's logon language	LANG	1
MANDT	R/3 System	Current client	CLNT	3
MODNO	R/3 System	Index of the external sessions	CHAR	1
OPSYS	R/3 System	Operating system of application server	CHAR	10
SAPRL	R/3 System	R/3 Release in use	CHAR	4
SYSID	R/3 System	Name of the R/3 System	CHAR	8
UNAME	R/3 System	Username of current user	CHAR	12
SUBRC	Return code	Return code following an ABAP statement	INT4	10
CUCOL	Screens	Horizontal cursor position in PAI (column)	INT4	10
CUROW	Screens	Vertical cursor position in PAI (line)	INT4	10
DATAR	Screens	'X' = Displays user input	CHAR	1

Field Name	Use	Description	Type	Len
LOOPC	Screens	Number of LOOP lines visible in the table	INT4	10
PFKEY	Screens	Current GUI status	CHAR	20
SCOLS	Screens	Number of columns in window	INT4	10
SROWS	Screens	Number of lines in window	INT4	10
STEPL	Screens	Index of current table line	INT4	10
TITLE	Screens	Text in the title bar	CHAR	70
UCOMM	Screens	Function code that triggered PAI	CHAR	70
SLSET	Selection screens	Variant name	CHAR	14
FDPOS	Strings	Offset in a string after a search hit	INT4	10

Appendix A2 - System Fields (by field name)

SAP systems maintain a runtime array called `SY` that contains a lot of data about the system and the program that's running. You can get the values of these fields (and assign values to some of them) by referring to them as `SY-XXXXX` where `XXXXX` is the field name. For example the user's login name is stored in `SY-UNAME`. R/3 releases 2.x and 3.x have the same system fields while releases 4.0 and 4.x each have slightly different lists of system fields. The fields in this list are valid for all releases through 4.6 unless otherwise identified. 78 of the original system fields are obsolete or for internal use only; those are not shown here.

Field Name	Use	Description	Type	Len
ABCDE	Constants	Alphabet (A,B,C,...)	CHAR	26
BATCH	Background processing	'X' = program is running in the background	CHAR	1
BINPT	Batch input	'X' = program is running in the background	CHAR	1
CALLD	ABAP program	Call mode of the ABAP program	CHAR	1
CALLR	Printing lists	ID for print dialog function	CHAR	8
COLNO	Creating lists	Current list column	INT4	10
CPAGE	Processing lists	Current page number	INT4	10
CPROG	ABAP program	Program that called the current external procedure	CHAR	40
CUCOL	Screens	Horizontal cursor position in PAI (column)	INT4	10
CUROW	Screens	Vertical cursor position in PAI (line)	INT4	10
DATAR	Screens	'X' = Displays user input	CHAR	1
DATLO	Date and time	User's local date; releases 4.x only	DATS	8
DATUM	Date and time	Current application server date	DATS	8
DATUT	Date and time	Global UTC (GMT) date; release 4.0 only	DATS	8
DAYST	Date and time	'X' = summer (daylight saving) time	CHAR	1
DBCNT	Database access	Number of database rows processed	INT4	10
DBNAM	ABAP program	Logical database linked to the program	CHAR	20
DBSYS	R/3 System	Name of the central database system (e.g. Oracle, Informix, etc.)	CHAR	10
DYNGR	ABAP program	Screen group of the current screen	CHAR	4
DYNNR	ABAP program	Number of the current screen	CHAR	4
FDAYW	Date and time	Weekday in the factory calendar	INT1	3

Field Name	Use	Description	Type	Len
FDPOS	Strings	Offset in a string after a search hit	INT4	10
HOST	R/3 System	Name of application server	CHAR	8
INDEX	Loops	Current loop pass in DO and WHILE loop	INT4	10
LANGU	R/3 System	User's logon language	LANG	1
LDBPG	ABAP program	Logical database program for SY-DBNAM	CHAR	40
LILLI	Processing lists	List line selected	INT4	10
LINCT	Creating lists	Page length in a list (from REPORT)	INT4	10
LINNO	Creating lists	Current line	INT4	10
LINSZ	Creating lists	Line width in a list (from REPORT)	INT4	10
LISEL	Processing lists	Contents of the selected line	CHAR	255
LISTI	Processing lists	Index of the selected list (0=base, 1=detail 1 ...)	INT4	10
LOOPC	Screens	Number of LOOP lines visible in the table	INT4	10
LSIND	Processing lists	Index of the detail list (0=base, 1=detail 1 ...)	INT4	10
MACOL	Printing lists	Left columns from the SET MARGIN statement	INT4	10
MANDT	R/3 System	Current client	CLNT	3
MAROW	Printing lists	Top rows from the SET MARGIN statement	INT4	10
MODNO	R/3 System	Index of the external sessions	CHAR	1
MSGID	Messages	Message class(application area)	CHAR	20
MSGLI	Messages	Contents of the message line	CHAR	60
MSGNO	Messages	Message number	NUMC	3
MSGTY	Messages	Message type (A,E,I,S,W,X)	CHAR	1
MSGV1	Messages	Message variable	CHAR	50
MSGV2	Messages	Message variable	CHAR	50
MSGV3	Messages	Message variable	CHAR	50
MSGV4	Messages	Message variable	CHAR	50
OPSYS	R/3 System	Operating system of application server	CHAR	10
PAART	Printing lists	Print formatting	CHAR	16
PAGNO	Creating lists	Current page	INT4	10
PDEST	Printing lists	Output device	CHAR	4
PEXPI	Printing lists	Spool retention period	NUMC	1
PFKEY	Screens	Current GUI status	CHAR	20
PLIST	Printing lists	Name of spool request	CHAR	12
PRABT	Printing lists	Cover sheet: Department	CHAR	12
PRBIG	Printing lists	'X' = Selection cover sheet	CHAR	1
PRCOP	Printing lists	Number of copies	NUMC	3

Field Name	Use	Description	Type	Len
PRDSN	Printing lists	Name of the spool dataset	CHAR	6
PRIMM	Printing lists	'X' = Print immediately	CHAR	1
PRNEW	Printing lists	'X' = New spool request (list)	CHAR	1
PRREC	Printing lists	Recipient	CHAR	12
PRREL	Printing lists	'X' = Delete after printing	CHAR	1
PRTXT	Printing lists	Text for cover sheet	CHAR	68
REPID	ABAP program	Current main program	CHAR	40
RTITL	Printing lists	Program from which you are printing	CHAR	70
SAPRL	R/3 System	R/3 Release in use	CHAR	4
SCOLS	Screens	Number of columns in window	INT4	10
SLSET	Selection screens	Variant name	CHAR	14
SPONO	Printing lists	Spool number	NUMC	10
SROWS	Screens	Number of lines in window	INT4	10
STACO	List processing	Number of first displayed column	INT4	10
STARO	List processing	Number of first displayed line on this page	INT4	10
STEPL	Screens	Index of current table line	INT4	10
SUBRC	Return code	Return code following an ABAP statement	INT4	10
SYSID	R/3 System	Name of the R/3 System	CHAR	8
TABIX	Internal tables	Current line of an internal table	INT4	10
TCODE	ABAP program	Current transaction code	CHAR	20
TFILL	Internal tables	Current number of entries in internal table	INT4	10
TIMLO	Date and time	User's local time; releases 4.x only	TIMS	6
TIMUT	Date and time	Global UTC (GMT) time; release 4.0 only	TIMS	6
TITLE	Screens	Text in the title bar	CHAR	70
TLENG	Internal tables	Line width of an internal table	INT4	10
TOCCU	Internal tables	OCCURS parameter with internal tables	INT4	10
TSTLO	Date and time	Local date and time; release 4.0 only	NUMC	14
TSTUT	Date and time	UTC (GMT) date and time; release 4.0 only	NUMC	14
TVAR0	Creating lists	Text variable for ABAP text elements	CHAR	20
TVAR1	Creating lists	Text variable for ABAP text elements	CHAR	20
TVAR2	Creating lists	Text variable for ABAP text elements	CHAR	20
TVAR3	Creating lists	Text variable for ABAP text elements	CHAR	20
TVAR4	Creating lists	Text variable for ABAP text elements	CHAR	20
TVAR5	Creating lists	Text variable for ABAP text elements	CHAR	20
TVAR6	Creating lists	Text variable for ABAP text elements	CHAR	20

Field Name	Use	Description	Type	Len
TVAR7	Creating lists	Text variable for ABAP text elements	CHAR	20
TVAR8	Creating lists	Text variable for ABAP text elements	CHAR	20
TVAR9	Creating lists	Text variable for ABAP text elements	CHAR	20
TZONE	Date and time	Difference between local time and GMT in seconds	INT4	10
UCOMM	Screens	Function code that triggered PAI	CHAR	70
ULINE	Constants	Horizontal line with length 255	CHAR	255
UNAME	R/3 System	Username of current user	CHAR	12
UZEIT	Date and time	Current application server time	TIMS	6
VLINE	Constants	Vertical line (|)	CHAR	1
WTITL	Creating lists	'X' = standard page header	CHAR	1
ZONLO	Date and time	User's time zone; releases 4.x only	CHAR	6

Appendix B1 - Transaction Codes (by description)

This is a list of a few selected transaction codes that the ABAP programmer may find useful. SAP is de-emphasizing the use of transaction codes, preferring the use of menu paths instead. Some transaction codes may disappear in later releases.

TCode	Description
SE12	ABAP Dictionary Display
SE11	ABAP Dictionary Maintenance
ST22	ABAP dump analysis
SE38	ABAP Editor
SE37	ABAP Function Library
SQ02	ABAP Query - functional areas
SQ03	ABAP Query - user groups
SQ01	ABAP Query develop & execute
SE39	ABAP Split-screen Editor
SE30	ABAP Trace
S001	ABAP Workbench (CASE Menu)
F040	Archive Data
SU20	Authorization fields maintenance
SU21	Authorization object maintenance
SM36	Batch Job Definition
SM35	Batch Job Monitoring
SM37	Batch Job Overview (status)
SB01	Business Navigator - Component view
SB02	Business Navigator - Process view
SE01	Correction & Transport System (→SE09)
SE15	Data Dictionary Information System
DI04	Data elements: Repository Information (3.1 and below)
SD11	Data Modeler
SXDA	Data Transfer Workbench
SE14	Database Utility
ST11	Developer trace display
SE35	Dialog Modules

TCode	Description
SM12	Display Locks
SMX	Display own jobs
DI01	Domains: Repository Information (3.1 and below)
SMEN	Dynamic Menu
SDBE	Explain SQL
DI24	Field in Matchcode IDs: Repository Information (3.1 and below)
DI27	Field use (aggregates): Repository Information (3.1 and below)
DI03	Fields: Repository Information (3.1 and below)
SE73	Font Maintenance
SHDG	Global Values
DI12	Indexes: Repository Information (3.1 and below)
SM59	Install 3rd-party product
SE63	Language translation of field labels
DI22	Lock objects: Repository Information (3.1 and below)
SM01	Lock Transactions
SA01	Locking
ALDB	Logical database manager
SE36	Logical Databases
SE91	Maintain Messages
SE93	Maintain Transaction Codes
DI23	Matchcode IDs: Repository Information (3.1 and below)
DI20	Matchcode objects: Repository Information (3.1 and below)
SE41	Menu Painter
T100	Message ID List
SE80	Object Browser (Development Workbench)
DI06	Pools/Clusters: Repository Information (3.1 and below)
SM64	Raise Events
SHDB	Record a transaction to create a BDC
DI08	Relationships (Foreign Keys): Repository Information (3.1 and below)
SA38	Report Execution/Launch
SE84	Repository Information
SE71	SAPScript - layout set request
SE53	Screen Foreign Language Maintenance
SE51	Screen Painter
SM02	Send system messages

TCode	Description
SPAD	Spool Administrator
SP01	Spool Manager
ST05	SQL Trace
SM50	System Administration
SM51	System Administration
SM04	System Administration (User Overview)
SM21	System Log Viewer
S000	System Main Menu
ST01	System trace
SM31	Table maintainer
DI26	Table use (aggregates): Repository Information (3.1 and below)
SE16	Table view & maintain
SE17	Table viewer (former SE16)
SM30	Table views maintainer
DI02	Tables: Repository Information (3.1 and below)
SP12	TemSe (Spool data file) Administration
SP11	TemSe (Spool data file) Contents
SE32	Text Element Maintenance
STOR	Trace Analysis
STOM	Trace Option Settings
SE05	Transport Information
SE07	Transport System Status Display
SE03	Transport System Utilities
SM13	Update Requests
SU51	User address maintenance
SU50	User default maintenance
BIBS	User interface examples
SU01	User Master maintenance
SU54	User menu
SU52	User parameters maintenance
SU02	User Profiles maintenance
SU56	User's authorizations display
SU53	User's most recent authorization check
DI25	Views: Repository Information (3.1 and below)
SO01	Word processor
SE09	Workbench Organizer

Appendix B2 - Transaction Codes (by TransCode)

This is a list of a few selected transaction codes that the ABAP programmer may find useful. SAP is de-emphasizing the use of transaction codes, preferring the use of menu paths instead. Some transaction codes may disappear in later releases.

TCode	Description
ALDB	Logical database manager
BIBS	User interface examples
DI01	Domains: Repository Information (3.1 and below)
DI02	Tables: Repository Information (3.1 and below)
DI03	Fields: Repository Information (3.1 and below)
DI04	Data elements: Repository Information (3.1 and below)
DI06	Pools/Clusters: Repository Information (3.1 and below)
DI08	Relationships (Foreign Keys): Repository Information (3.1 and below)
DI12	Indexes: Repository Information (3.1 and below)
DI20	Matchcode objects: Repository Information (3.1 and below)
DI22	Lock objects: Repository Information (3.1 and below)
DI23	Matchcode IDs: Repository Information (3.1 and below)
DI24	Field in Matchcode IDs: Repository Information (3.1 and below)
DI25	Views: Repository Information (3.1 and below)
DI26	Table use (aggregates): Repository Information (3.1 and below)
DI27	Field use (aggregates): Repository Information (3.1 and below)
F040	Archive Data
S000	System Main Menu
S001	ABAP Workbench (CASE Menu)
SA01	Locking
SA38	Report Execution/Launch
SB01	Business Navigator - Component view
SB02	Business Navigator - Process view
SD11	Data Modeler
SDBE	Explain SQL
SE01	Correction & Transport System (→SE09)
SE03	Transport System Utilities

TCode	Description
SE05	Transport Information
SE07	Transport System Status Display
SE09	Workbench Organizer
SE11	ABAP Dictionary Maintenance
SE12	ABAP Dictionary Display
SE14	Database Utility
SE15	Data Dictionary Information System
SE16	Table view & maintain
SE17	Table viewer (former SE16)
SE30	ABAP Trace
SE32	Text Element Maintenance
SE35	Dialog Modules
SE36	Logical Databases
SE37	ABAP Function Library
SE38	ABAP Editor
SE39	ABAP Split-screen Editor
SE41	Menu Painter
SE51	Screen Painter
SE53	Screen Foreign Language Maintenance
SE63	Language translation of field labels
SE71	SAPScript - layout set request
SE73	Font Maintenance
SE80	Object Browser (Development Workbench)
SE84	Repository Information
SE91	Maintain Messages
SE93	Maintain Transaction Codes
SHDG	Global Values
SHDB	Record a transaction to create a BDC
SM01	Lock Transactions
SM02	Send system messages
SM04	System Administration (User Overview)
SM12	Display Locks
SM13	Update Requests
SM21	System Log Viewer
SM30	Table views maintainer
SM31	Table maintainer

TCode	Description
SM35	Batch Job Monitoring
SM36	Batch Job Definition
SM37	Batch Job Overview (status)
SM50	System Administration
SM51	System Administration
SM59	Install 3rd-party product
SM64	Raise Events
SMEN	Dynamic Menu
SMX	Display own jobs
SO01	Word processor
SP01	Spool Manager
SP11	TemSe (Spool data file) Contents
SP12	TemSe (Spool data file) Administration
SPAD	Spool Administrator
SQ01	ABAP Query develop & execute
SQ02	ABAP Query - functional areas
SQ03	ABAP Query - user groups
ST01	System trace
ST05	SQL Trace
ST11	Developer trace display
ST22	ABAP dump analysis
STOM	Trace Option Settings
STOR	Trace Analysis
SU01	User Master maintenance
SU02	User Profiles maintenance
SU20	Authorization fields maintenance
SU21	Authorization object maintenance
SU50	User default maintenance
SU51	User address maintenance
SU52	User parameters maintenance
SU53	User's most recent authorization check
SU54	User menu
SU56	User's authorizations display
SXDA	Data Transfer Workbench
T100	Message ID List

Appendix C1 - System Tables (by description)

This is a list of a few selected tables that the ABAP programmer may find useful. View tables in /SM16; maintain them in /SM30 or /SM31.

Tablename	Description
TBRGT	Authorization object & group descriptions
TSYST	Available SAP Systems
TCP01	Character Set - SAP
TCP05	Character Set Mfr
TCP00	Character Sets
TCP02	Character Sets - Edit
TCP03	Character Sets - Edit
TCP07	Character Sets - Edit
T004T	Chart of Account Names
T000	Clients (in German: Mandt)
TSE05	Code Templates
T001	Companies
T005T	Country Names
T005X	Country numeric & date formats
Z..	Customer-created tables
Y..	Customer-created tables
DDNT	Data Dictionary Objects
DD02V	Data dictionary tables
DD03L	Data dictionary tables & fields
TDEVC	Development Classes
TDCT	Dialog module list
T015Z	Digits and numbers in text
T003	Document Types
T003T	Document Types, Text
TFACS	Factory Calendar display
TFDIR	Function Modules
P9..	HR module tables
T002T	Language key, text
T100	Messages

Tablename	Description
TFAWF	Modifiable fields
T015M	Month Names
TADIR	Object Catalog
T100W	Plants
T022D	Print Control for Device Type
TSP1D	Printer - Types of Formatting
TFO06	Printer Barcodes
TSP0A	Printer Device Type
TFO03	Printer Fonts
TSP06	Printer Formatting for Device Types
TSP03	Printer Output Device
TSP03C	Printer Output Device
TSP08	Printer Page Formats
TRCL	Program Classes
TRCL	Program classes
TRDIR	Programs
TASYS	Recipient (Target) system(s) of a transport
TRESE	Reserved words
T185F	Screen control function codes
TAORA	Technical Settings of Tables
TGORA	Technical Settings of Tables
TSTCT	Transaction Codes & descriptions
T006	Units of measure
USR03	User address data
USR02	User login data
USR04	User master authorizations
USR01	User master record at runtime
USR05	User SPA/GPA parameter values
TVARE	Variant Values
VARI	Variants
DD02T	View of all tables
TVDIR	Views

Appendix C2 - System Tables (by table name)

This is a list of a few selected tables that the ABAP programmer may find useful. View tables in `/SM16`; maintain them in `/SM30` or `/SM31`.

Tablename	Description
DD02T	View of all tables
DD02V	Data dictionary tables
DD03L	Data dictionary tables & fields
DDNT	Data Dictionary Objects
P9..	HR module tables
T000	Clients (in German: Mandt)
T001	Companies
T002T	Language key, text
T003	Document Types
T003T	Document Types, Text
T004T	Chart of Account Names
T005T	Country Names
T005X	Country numeric & date formats
T006	Units of measure
T015M	Month Names
T015Z	Digits and numbers in text
T022D	Print Control for Device Type
T100	Messages
T100W	Plants
T185F	Screen control function codes
TADIR	Object Catalog
TAORA	Technical Settings of Tables
TASYS	Recipient (Target) system(s) of a transport
TBRGT	Authorization object & group descriptions
TCP00	Character Sets
TCP01	Character Set - SAP
TCP02	Character Sets - Edit
TCP03	Character Sets - Edit
TCP05	Character Set Mfr

Tablename	Description
TCP07	Character Sets - Edit
TDCT	Dialog module list
TDEVC	Development Classes
TFACS	Factory Calendar display
TFAWF	Modifiable fields
TFDIR	Function Modules
TFO03	Printer Fonts
TFO06	Printer Barcodes
TGORA	Technical Settings of Tables
TRCL	Program Classes
TRCL	Program classes
TRDIR	Programs
TRESE	Reserved words
TSE05	Code Templates
TSP03	Printer Output Device
TSP03C	Printer Output Device
TSP06	Printer Formatting for Device Types
TSP08	Printer Page Formats
TSP0A	Printer Device Type
TSP1D	Printer - Types of Formatting
TSTCT	Transaction Codes & descriptions
TSYST	Available SAP Systems
TVARE	Variant Values
TVDIR	Views
USR01	User master record at runtime
USR02	User login data
USR03	User address data
USR04	User master authorizations
USR05	User SPA/GPA parameter values
VARI	Variants
Y..	Customer-created tables
Z..	Customer-created tables

Appendix D – Type Conversions

SAP automatically converts field types when it can, as described in the table below. When making the assignment `Target` `Source` if the target is shorter than the source the value may be truncated and it may be flagged by or filled with asterisks. If the target is longer it may be filled with blanks or zeros. The table indicates whether truncation is on the left or right end of the target, and how the value is justified in the target.

Column Title	Value	Description
Short Tgt Trunc If the target is shorter than required by the source:	`na`	Not applicable
	`L`	Truncates left end of target
	`L*`	Truncates the left end and appends a '*' at the left end
	`R`	Truncates right end of target
	`Rnd*`	Rounds to fit; fills with '*' when target is too short for rounded value
	`OFE`	Overflow error
Long Tgt Fill If the target is longer than required by the source:	`L'`	Fills left end of target with spaces
	`L0`	Fills left end of target with zeros
	`L0F`	Fills left end of target with zeros or hexadecimal `F`
	`R'`	Fills right end of target with spaces
	`R0`	Fills right end of target with zeros
Justification	`L`	Value is Left justified in target
	`R`	Value is Right justified in target
Negative Source If the source contains a negative value, the target will receive it with the following indication:	`na`	Not applicable
	`No`	No sign transferred
	`L±`	Plus sign or minus sign on left end of target
	`L-`	Minus sign on left end of target if needed
	`R±`	Plus sign or minus sign on right end of target
	`R-`	Space or minus sign on right end of target
	`R'`	Space on right end of target
	`2s`	Two's complement (negative hexadecimal)
	`D-`	Negative source values are assigned as 01/01/0001

Column Title	Value	Description
	T-	Number of seconds before midnight
Conversion Details	→ in any column	Refer to Conversion Details for more information

Target = Source	Short Tgt Trunc	Long Tgt Fill	Justification	Negative	Conversion Details
C = C	L	R''	L	na	Assigned.
C = D	R	R''	L	na	Converted to YYYYMMDD format and assigned.
C = F	Rnd*	L''	R	L±	Assigned in the format: mantissa E exponent where 1.0 ≤ mantissa <10 and the exponent contains at least two digits.
C = I	L*	L''	R	R-	Assigned with trailing minus sign or blank.
C = N	R	R''	L	na	Assignment, no conversion; leading zeros remain.
C = P	L*	L''	R	R-	Unpacked. Assigned with decimal-point if needed and trailing minus sign or blank. See UNPACK for zero-filled long target.
C = STRING	L	R''	L	na	Assigned.
C = T	R	R''	L	na	Converted to HHMMSS format from storage format (number of seconds since midnight) then assigned.
C = X	R	R''	L	na	Converted to hexadecimal representation (1-9,A-F) and assigned.
C = XSTRING	R	R''	L	na	Converted to hexadecimal representation (1-9,A-F) and assigned.
D = C	R	L''	R	na	First 8 characters must be digits which are interpreted as YYYYMMDD.
D = D	-	-	-	na	Assigned.
D = F	→	-	-	D-	Rounded to an integer modulo 3,652,426.. Interpreted as the number of days since 01/01/0001 and converted to date format. Negative source values are assigned as 01/01/0001.
D = I	→	-	-	D-	Interpreted as the number of days since 01/01/0001 modulo 3,652,426 (01/01/10,000), and converted to date format. Negative source values are converted as 01/01/0001.
D = N	R	L''	R	na	First 8 characters must be digits which are interpreted as YYYYMMDD.

Target = Source	Short Tgt Trunc	Long Tgt Fill	Justification	Negative	Conversion Details
D = NOT D	-	-	-	na	See CONVERT for conversion to nines-complement date.
D = P	→	-		D-	Unpacked & rounded to an integer, Interpreted as the number of days since 01/01/0001 modulo 3,652,061 (01/01/10,000), and converted to date format. Negative source values are assigned as 01/01/0001.
D = STRING	R	L''	R	na	First 8 characters must be digits which are interpreted as YYYYMMDD.
D = T	-	-	-	-	Invalid; program dump.
D = X	→	-	-	D-	Interpreted as the number of days since 01/01/0001 modulo 3,652,061 (01/01/10,000), and converted to date format.
D = XSTRING	→	-	-	D-	Interpreted as the number of days since 01/01/0001 modulo 3,652,061 (01/01/10,000), and converted to date format.
F = C	→	R0	-	L±	Resolution is limited to about 15 digits. Source must contain a floating point number in any valid representation: 3, 3.14, -3, -3.14, 3E+5, -3E5, 3.14E-5, -3.14E-5
F = D	-	R0	-	na	Assigned as a floating point number equal to the number of days since 01/01/0001.
F = F	-	R0	-	L±	Assigned without conversion.
F = I	-	R0	-	L±	Assigned as a floating point number.
F = N	→	R0	-	L+	Resolution is limited to about 15 digits. Source must contain a floating point number in any valid representation: 3, 3.14, -3, -3.14, 3E+5, -3E5, 3.14E-5, -3.14E-5
F = P	→	R0	-	L±	Assigned as a floating point number.
F = STRING	→	R0	-	L±	Resolution is limited to about 15 digits. Source must contain a floating point number in any valid representation: 3, 3.14, -3, -3.14, 3E+5, -3E5, 3.14E-5, -3.14E-5
F = T	-	R0	-	L+	The number of seconds since midnight are assigned as a floating point number.
F = X	-	R0	-	L±	The highest four bytes of Source are converted from hexadecimal to decimal floating point and assigned.
F = XSTRING	-	R0	-	L±	Up to four bytes of Source are converted from hexadecimal to decimal floating point and assigned.

Target = Source	Short Tgt Trunc	Long Tgt Fill	Justification	Negative	Conversion Details
I = C	OFE	L''	R	R-	Source must contain a numeric string and may include a sign and/or a leading or trailing decimal point.
I = D	-	L''	R	R''	Assigned as an integer equal to the number of days since 01/01/0001.
I = F	OFE	L''	R	R-	Rounded to an integer and assigned.
I = I	-	-	-	R-	Assigned.
I = N	OFE	L''	R	R''	Assigned without leading zeros.
I = P	OFE	L''	R	R-	Rounded to an integer and assigned.
I = STRING	OFE	L''	R	R-	Source must contain a numeric string and may include a sign and/or a leading or training decimal point.
I = T	-	L''	R	R''	The number of seconds since midnight are assigned as an integer.
I = X	-	L''	R	R-	The highest four bytes of Source are converted from hexadecimal to decimal integer and assigned.
I = XSTRING	-	L''	R	R-	Up to four bytes of Source are converted from hexadecimal to decimal integer and assigned.
N = C	L	L0	R	No	Digits assigned, non-digits are ignored.
N = D	L	R0	L	na	Converted to YYYYMMDD format and assigned.
N = F	L	L0	R	No	Rounded, then assigned without sign.
N = I	L	L0	R	No	Assigned without sign.
N = N	L	L0	R	na	Assigned.
N = P	L	L0	R	No	Rounded to an integer and assigned without sign.
N = STRING	L	L0	R	No	Digits assigned, non-digits are ignored.
N = T	R	R0	L	na	Converted to HHMMSS format from storage format (number of seconds since midnight) then assigned.
N = X	R	R''	L	na	The highest four bytes of Source are converted from hexadecimal to decimal and assigned.
N = XSTRING	R	R''	L	na	Up to four bytes of Source are converted from hexadecimal to decimal and assigned.
P = C	OFE	L''	R	R-	Packed and assigned; Source must contain a numeric string and may include a sign and/or a decimal point.
P = D	OFE	L''	R	R-	Converted to an integer equal to the number of days since 01/01/0001, packed and assigned.

Target = Source	Short Tgt Trunc	Long Tgt Fill	Justification	Negative	Conversion Details
P = F	OFE	L''	R	R-	Source is rounded to an integer, packed and assigned.
P = I	OFE	L''	R	R-	Packed and assigned.
P = N	OFE	L''	R	R''	Packed and assigned as a positive number.
P = P	OFE	L''	R	R-	Assigned.
P = STRING	OFE	L''	R	R-	Packed and assigned: Source must contain a numeric string and may include a sign and/or a decimal point.
P = T	OFE	L''	R	R''	The number of seconds since midnight are packed and assigned.
P = X	OFE	L''	R	R-	The highest four bytes of Source are converted from hexadecimal to decimal, packed and assigned.
P = XSTRING	OFE	L''	R	R-	Up to four bytes of Source are converted from hexadecimal to decimal, packed and assigned.
STRING = C	L	R''	L	na	Assigned.
STRING = D	R	R''	L	na	Converted to YYYYMMDD format and assigned.
STRING = F	Rnd*	L''	R	L±	Assigned in the format: mantissa E exponent where 1.0 ≤ mantissa <10 and the exponent contains at least two digits.
STRING = I	L*	L''	R	R-	Assigned with trailing minus sign or blank.
STRING = N	R	R''	L	na	Assignment, no conversion; leading zeros remain.
STRING = P	L*	L''	R	R-	Unpacked. Assigned with decimal-point if needed and trailing minus sign or blank. See UNPACK for zero-filled long target.
STRING = T	R	R''	L	na	Converted to HHMMSS format from storage format (number of seconds since midnight) then assigned.
STRING = X	R	R''	L	na	Converted to hexadecimal representation (1-9,A-F) and assigned.
STRING = XSTRING	R	R''	L	na	Converted to hexadecimal representation (1-9,A-F) and assigned.
T = C	R	R0	L	na	First 6 characters must be digits which are interpreted as HHMMSS.
T = D	-	-	-	-	Invalid: program dump.

Target = Source	Short Tgt Trunc	Long Tgt Fill	Justification	Negative	Conversion Details
T = F	na	-	-	T-	Rounded to an integer and interpreted as the number of seconds since midnight modulo 86,400.
T = I	na	-	-	T-	Interpreted as the number of seconds since midnight modulo 86,400.
T = N	R	R0	L	na	First 6 characters must be digits which are interpreted as HHMMSS.
T = P	na	-	-	T-	Unpacked. Rounded to an integer and interpreted as the number of seconds since midnight modulo 86,400.
T = STRING	R	R0	L	na	First 6 characters must be digits which are interpreted as HHMMSS.
T = T	-	-	-	na	Assigned
T = X	na	-	-	T-	Interpreted as the number of seconds since midnight modulo 86,400.
T = XSTRING	na	-	-	T-	Interpreted as the number of seconds since midnight modulo 86,400.
X = C	R	R0	L	na	Source must contain a hexadecimal string (0-9,A-F where A-F are capitalized) with no sign. Conversion ceases at the first non-hex character.
X = D	L	R0	L	na	Converted to the number of days since 01/01/0001 in hexadecimal and assigned.
X = F	L	L0F	R	2s	Rounded, converted to hexadecimal, then assigned.
X = I	L	L0F	R	2s	Converted to hexadecimal and assigned.
X = N	L	L0	R	na	Converted to hexadecimal and assigned.
X = P	LL	L0F	R	2s	Unpacked, rounded to an integer, converted to hexadecimal and assigned.
X = STRING	R	R0	L	na	Assigned. Source must contain a hexadecimal string (0-9,A-F where A-F are capitalized) with no sign. Conversion ceases at the first non-hex character.
X = T	L	L0	R	na	The number of seconds since midnight are converted to hexadecimal and assigned.
X = X	R	R0	L	na	Assigned.
X = XSTRING	R	R0	L	na	Assigned.
XSTRING = C	R	R0	L	na	Assigned. Source must contain a hexadecimal string (0-9,A-F where A-F are capitalized) with no sign. Conversion ceases at the first non-hex character.
XSTRING = D	L	R0	L	na	Converted to the number of days since 01/01/0001 in hexadecimal and assigned.

Target = Source	Short Tgt Trunc	Long Tgt Fill	Justification	Negative	Conversion Details
XSTRING = F	L	L0F	R	2s	Rounded, converted to hexadecimal, then assigned.
XSTRING = I	L	L0F	R	2s	Converted to hexadecimal and assigned. Only the highest 4 bytes are converted.
XSTRING = N	L	L0	R	na	Converted to hexadecimal and assigned.
XSTRING = P	LL	L0F	R	2s	Unpacked, rounded to an integer, converted to hexadecimal and assigned.
XSTRING = STRING	R	R0	L	na	Assigned. Source must contain a hexadecimal string (0-9,A-F where A-F are capitalized) with no sign. Conversion ceases at the first non-hex character.
XSTRING = T	L	L0	R	na	The number of seconds since midnight are converted to hexadecimal and assigned.
XSTRING = X	R	R0	L	na	Assigned.

Appendix E – Command-line commands

This is a list of a few selected commands that the ABAP programmer may find useful. Type the command in the command field and press `<Enter>`.

Command	Description
`/BDEL`	BDC – Delete current transaction from session
`/BDEND`	BDC – Terminate processing and mark as incorrect
`/BDDA`	BDC – Change to "Display All" mode
`/BDDE`	BDC – Change to "Display Errors" mode
`/&HD`	Authorizations {System {User Profile {Hold Data
`/I`	Close the current session
`/Otcode`	Create a new session and launch transaction `tcode`
`/H`	Debug (Hobbel) Mode
`/Ntcode`	Depart current transaction and launch transaction `tcode`
`/N`	End current transaction; end debug mode
`/NEND`	Logoff
`/O`	System Session Overview

Appendix F –Utility Programs

This is a list of a few selected utility programs that the ABAP programmer may find useful. Execute programs in transaction SA38 or SE38.

Program	Description
`ABAPDOCU`	Programs showing examples of displaying, execution and debugging.
`D10SINF`	Information about source code
`RSDYNL10`	Lists screens (dynpros) associated with a report
`RSHOWTIM`	ABAP tips & tricks
`RSANAL00`	Analysis of a program: Variables Subroutines Analysis messages Conversions Programs / transactions External tables Statistics Source code
`RSINCL00`	Documents a program by listing its: Source code Expanded `INCLUDE` lines `INCLUDE` reference list `MODULE` reference list `FORM/PERFORM` reference list Function Module reference list Dialog Module reference list `SELECT` reference list `SELECT SINGLE` reference list `READ TABLE` reference list `LOOP AT` reference list `MODIFY` reference list `DELETE` reference list `MESSAGE` reference list `SCREEN` reference list `PF-STATUS` reference list

Program	Description
	SET/GET params reference list
	Field reference list
RSPARAM	Batch jobs Profile Report
RSPO0041	Spool Status Report
RSPFPAR	System parameters
RSSDOCTB	Table structure and links
SAPMACO	Create matchcode object
SAPMSOS0	UNIX command line

Appendix G - Examples

G.1 - Batch Data Communications (BDC)

The structure of the table BCDDATA is:

`Program(40)`	Module pool (program name)
`Dynpro(4) TYPE N`	Screen number
`Dynbegin(1)`	Starting a screen = `'X'`,`' '` otherwise
`fnam(132)`	Field name
`fval(132)`	Field value to assign

The internal table (`bdctab` in this example) has two types of records: (1) a screen is started by entering values in the first three fields of `bdctab`, and (2) screen fields are filled and controls are activated for that screen by entering values on just the last two fields of `bdctab` (the screen number is implicit until the next screen-starting record is encountered). Function keys are activated by entering `BDC_OKCODE` in `fnam` and `'/n'` in `fval` where `n` is the function key number (for example "Save" is `F11`, so `fval = '/11'`). The records in `bdctab` specify a sequence of screen actions to complete the transaction; in the following example the populated `bdctab` looks like:

Program	Dyn pro	Dyn begin	fnam	fval	(Notes)
`'SAPMFD02'`	`'0106'`	`'X'`			*first screen*
			`'RFD02-kunnr'`	`rec-kunnr`	*vendor #*
			`'RFD02-D0110'`	`'X'`	*"Address" box*
`'SAPMFD02'`	`'0110'`	`'X'`			*second screen*
			`'knal-telf1'`	`rec-telf1`	*phone number*
			`'knal-name2'`	`name2`	*test flag*
			`'BDC_OKCODE'`	`UPDA`	*Update (Save)*

Find the program name and screen number from {System {Status while in the screen. Find the table and field name by highlighting the field, clicking on the [Help or [? button and selecting [Technical_data from the help screen.

The transaction may follow different paths interactively and in batch operation after the BDC table has been populated. If your apparently valid process breaks, execute the program with `displaymode` set to 'A' (all) to find out if screens appear that you didn't see interactively. You may see one or more modal dialog boxes labeled "Coding Block" which may be blank or contain some fields. You'll need to add those screens to your BDC table and may need to use some of the fields. The program name and screen number may be found on the message line (typically at the bottom left of your screen), and may be obscured by the modal dialog box; move the dialog box up to read the necessary information.

Other commands available in the BDC table are:

fnam	fval	(Notes)
fnam(n)	<val>	Identifies line number n of the named field in a multi-line block of a form.
BDC_CURSOR	fnam(n)	Moves cursor to the named screen field in line number n (multi-line form).
BDC_OKCODE	'/nn'	Press Function Key nn.
BDC_OKCODE	'/0' (zero)	<Enter> (This works but I've found no documentation for it).
BDC_OKCODE	'/8'	Continue.
BDC_OKCODE	'/11'	Post.
BDC_OKCODE	'CS'	F2 = double-click; "Cursor-Select" (replaces PICK).
BDC_OKCODE	'PICK'	F2 = double-click; select (replaced by CS).
BDC_OKCODE	'BACK'	F3; return to previous screen (green left arrow).
BDC_OKCODE	'%EX'	Depart this process (yellow up arrow).
BDC_OKCODE	'RW'	Cancel (Rollback Work).
BDC_OKCODE	'P--'	Up to top of list.
BDC_OKCODE	'P-'	Page up.
BDC_OKCODE	'P+'	Page down.
BDC_OKCODE	'P++'	Down to end of list.
BDC_OKCODE	'PRI'	Print.
BDC_OKCODE	tcode	Call the named transaction.

The following program illustrates how to import data from a sequential file and use the BDC mechanism to validate and insert the data in the proper SAP tables.

```
REPORT Z_BDC_01.
TABLES: kna1.
PARAMETERS:  name LIKE apqi-groupid DEFAULT SY-UNAME,
             infile(32) DEFAULT 'myinputfile'  LOWER CASE,
             outfile(32) DEFAULT 'myoutputfile' LOWER CASE.
DATA:  bdctab LIKE bdcdata OCCURS 7 WITH HEADER LINE,
       name2 LIKE kna1-name2 VALUE 'BDC-26',
       BEGIN OF rec,
         kunnr LIKE kna1-kunnr,
```

```
        telf1 LIKE kna1-telf1,
      END OF rec.
INITIALIZATION.
WRITE: SY-DATUM TO name2+11,
       SY-UZEIT TO name2+22.
START-OF-SELECTION.
OPEN DATASET infile FOR INPUT IN TEXT MODE.
IF SY-SUBRC NE 0.
  WRITE: / 'Failed to open', infile.
  EXIT.
ENDIF.
DELETE DATASET outfile.
IF SY-SUBRC NE 0.
  WRITE: / 'Failed to delete', outfile.
  EXIT.
ENDIF.
OPEN DATASET outfile FOR APPENDING IN TEXT MODE.
IF SY-SUBRC NE 0.
  WRITE: / 'Failed to open', outfile.
  EXIT.
ENDIF.
CALL FUNCTION 'BDC_OPEN_GROUP'
  EXPORTING
    CLIENT = SY-MANDT
    GROUP  = name
    USER   = name.      "Authorization check is on this field
WRITE: / 'Opening BDC session', SY-MANDT, name,
       SY-UZEIT, 'Return =', SY-SUBRC.
DO.
  READ DATASET infile INTO rec. "get a record from the file
  IF SY-SUBRC NE 0. EXIT. ENDIF.  "end of file
  SELECT SINGLE * FROM kna1 WHERE kunnr = rec-kunnr.
  IF SY-SUBRC = 0.
    WRITE: / rec-kunnr, rec-telf1.
    PERFORM generate_bdc_data. "load the data into bdctab
    CALL FUNCTION 'BDC_INSERT "execute the import function
        EXPORTING TCODE = 'FD02'
        TABLES DYNPROTAB = bdctab.
    WRITE: 'Return =', SY-SUBRC.
  ELSE.
    WRITE: / 'Customer not found:', rec-kunnr.
    TRANSFER rec TO outfile. "write unknown customer out
  ENDIF.
ENDDO.
CALL FUNCTION 'BDC_CLOSE_GROUP'.
WRITE: / 'Closing BDC session', SY-MANDT, NAME,
       SY-UZEIT, 'Return =', SY-SUBRC.
END-OF-SELECTION.
CLOSE DATASET outfile.
CLOSE DATASET infile.
FORM generate_bdc_data.
  WRITE: / 'in zc032601'.
```

```
  REFRESH bdctab. "empty the table
  PERFORM generate_screen  USING 'SAPMF02D' '0106'.
  PERFORM generate_entry   USING 'RF02D-kunnr' rec-kunnr.
  PERFORM generate_entry   USING 'RF02D-D0110' 'X'.
  PERFORM generate_screen  USING 'SAPMF02D' '0110'.
  PERFORM generate_entry   USING 'kna1-telf1' rec-telf1.
  PERFORM generate_entry   USING 'kna1-name2' name2.
  PERFORM generate_entry   USING 'BDC_OKCODE' 'UPDA'.
ENDFORM.
FORM generate_screen USING pname sname.
  CLEAR bdctab.              "empty the header line
  bdctab-program = pname.    "the program name
  bdctab-dynpro  = sname.    "the screen number
  bdctab-dynbegin = 'X'.     "select it
  APPEND bdctab.             "append the header to the table
ENDFORM.
FORM generate_entry USING fname fvalue.
  CLEAR bdctab.              "empty the header line
  bdctab-fnam = fname.       "entry field name
  bdctab-fval = fvalue.      "entry value
  APPEND bdctab.             "append the header to the table
ENDFORM.
* END OF REPORT
```

G.2 - **FIELD-SYMBOLS, ASSIGN**

Example of string concatenation in Release 2.2 (which doesn't have the CONCATENATE command).

```
PARAMETERS:   lsource(80)   DEFAULT 'lstring',
              rsource(80)   DEFAULT 'rstring',
              no_gaps       DEFAULT''.
DATA:  p(3) TYPE I,              "runtime offset
       w(3) TYPE I,              "runtime width
       target(80).               "target string
FIELD-SYMBOLS: <fs>.
p = STRLEN(lsource).
IF no_gaps = 'X'.
  p = STRLEN(lsource).
ELSE.
  p = STRLEN(lsource) + 1.
ENDIF.
w = STRLEN(rsource).
target = lsource.
* Point <fs> to the substring of target
ASSIGN target+p(w) TO <fs>.
* Assign the right string to the substring
<fs> = rsource.
WRITE: / 'Left =',       lsource,
       ', Right =', rsource,
       ', Result =', target.
```

```
→ Left = lstring, Right = rstring, Result = lstring rstring
```

Indirect addressing Example:

```
DATA: f(5),
      fname VALUE 'Bright Star',
      lname VALUE' Productions'.
FIELD-SYMBOLS <fs>.
ASSIGN (f) TO <fs>.
f = 'fname'.
WRITE / <fs>.
f = 'lname'.
WRITE <fs>.
→  Bright Star Productions
```

G.3 - Logical Database Processing

This is a description of how the `GET` event works. The more mnemonic syntactic construct would be `GOT`, rather than `GET`. If you use `DEFINING DATABASE ldb` in the header statement of your `REPORT`, then SAP starts an internal program when your `START-OF-SELECTION` event completes. That program (named `SAPDBldb` if the logical database is called `ldb`) reads every record in the logical database in hierarchical order. Every time a new table record is available, `SAPDBldb` looks for a `GET` event on that table in your program and triggers it if it exists. Every time a current table record is about to change (that is, all the lower level linked records have been processed), `SAPDBldb` looks for a `GET LATE` event on that table in your program and triggers it if it exists. The code blocks in those events can include `SELECT` and any other `ABAP` commands. These code blocks may trigger yet other events (such as `TOP-OF-PAGE`) which must be processed before the `GET` event can complete. For example, assume `ldb` has three tables linked as follows:

```
ldb STRUCTURE:
  header
    |
    +—subheader
          |
          +—detail
```

```
*===================================
REPORT DEFINING DATABASE ldb.
TABLES: header, detail.    "Note that subheader is missing.
SELECT-OPTIONS...
PARAMETERS:...
DATA:...
INITIALIZATION.
  ...
AT SCREEN SELECTION OUTPUT.
```

```
  ...
START OF SELECTION.
  ...
GET header.             "The next header record is available.
  "header processing code block...
GET LATE header.        "The current header record is about to change.
  "header end-of-record processing code block...
GET detail.             "The next detail record is available.
  "detail processing code block...
END-OF-SELECTION.
  ...
TOP-OF-PAGE.
  ...
FORM...
  ...
ENDFORM.
*(EOF)
*====================================
```

```
(DATA EXAMPLE:)                  TRIGGERED EVENTS
-----------------------------------------------------------------
                                 INITIALIZATION
                                 AT SELECTION-SCREEN...
                                 START-OF-SELECTION
header 1                         GET header 1
  subhead 1,1
    detail 1,1,1                 GET detail 1,1,1
    detail 1,1,2                 GET detail 1,1,2
    detail 1,1,3                 GET detail 1,1,3
  subhead 1,2
    detail 1,2,1                 GET detail 1,2,1
    detail 1,2,2                 GET detail 1,2,2
    detail 1,2,3                 GET detail 1,2,3
                                 GET LATE header 1
header 2                         GET header 2
  subhead 2,1
  subhead 2,2
     detail   2,2,1              GET detail 2,2,1
     detail   2,2,2              GET detail 2,2,2
     detail   2,2,3              GET detail 2,2,3
                                 GET LATE header 2
header 3                         GET header 3
  subhead 3,2                    GET LATE header 3
header 4                         GET header 4
  subhead 4,1
    detail 4,1,1                 GET detail 4,1
                                 GET LATE header 4
header 5                         GET header 5
                                 END-OF-SELECTION
```

In this example, there is an intervening table (`subheader`) that was not included in the `TABLES` statement; the `ldb` reader still marches through all its records (just to get to the `detail` records) but your program has no access to them.

G.4 - Pagination

Here's a brief program and its output that demonstrates some of the page-formatting features in ABAP.

```
REPORT ZDB_TEST LINE-COUNT 10(3) LINE-SIZE 62.
TOP-OF-PAGE.
  WRITE / 'Writing top-of-page'.
  BACK.
  WRITE 25 'Second top string'.
END-OF-PAGE.
  WRITE: 'Writing end-of-page for page', SY-PAGNO NO-GAP.
  WRITE / 'eop line 2'.
  WRITE / 'eop line 3'.
  BACK. "to the top of the list
  WRITE 35 'BACK from the EOP block'.
START-OF-SELECTION.
  DO.
    IF SY-INDEX GT 10.
        WRITE: / 'End of list;',
                 'notice that End-of-Page is not printed.'.
        EXIT.
    ENDIF.
    IF SY-INDEX EQ 3. BACK. ENDIF.
    IF SY-INDEX EQ 9.
        WRITE: / 'Forced NEW-PAGE;',
                 'notice that End-of-Page is not printed.'.
        NEW-PAGE.
    ENDIF.
    WRITE: / 'Writing line', SY-INDEX, SY-LINNO.
  ENDDO.
```

```
05.07.2001                    zdbtest                        1
--------------------------------------------------------------
Writing top-of-page     Second top string
Writing line        3      4      BACK from the EOP block
Writing line        4      5
Writing line        5      6
Writing line        6      7
Writing end-of-page for page      1
eop line 2
eop line 3
05.07.2001                    zdbtest                        2
--------------------------------------------------------------
Writing top-of-page     Second top string
```

```
Writing line        7      4
Writing line        8      5
Forced NEW-PAGE; notice that End-of-Page is not printed.



05.07.2001                    zdbtest                       3
--------------------------------------------------------------
Writing top-of-page     Second top string
Writing line        9      4
Writing line        10     5

End of list; notice that End-of-Page is not printed.
```

G.5 – Work Area

Here's an illustration of producing identical results from different types of work areas.

```
TABLES: kna1, *kna1.
DATA:  BEGIN OF wkna1.
          INCLUDE STRUCTURE kna1.
DATA:  END OF kna1.
DATA:  BEGIN OF ikna1 OCCURS 12.
          INCLUDE STRUCTURE kna1.
DATA: END OF ina1.

*Demonstrate implicit header line of table
WRITE 'kna1'.
SELECT * FROM kna1 WHERE KUNNR LT '0000100012'.
  WRITE: / kna1-kunnr.
ENDSELECT.
BACK.

*Demonstrate "star" header line of table
WRITE 15 '*kna1'.
SELECT * FROM kna1 INTO *kna1 WHERE KUNNR LT '0000100012'.
  WRITE: /15 *kna1-kunnr.
ENDSELECT.
BACK.

*Demonstrate explicitly-defines header line
WRITE 30 'wkna1'.
SELECT * FROM kna1 INTO wkna1 WHERE KUNNR LT '0000100012'.
  WRITE: /30 wkna1-kunnr.
ENDSELECT.
BACK.

*Demonstrate itab header line
WRITE 'ikna1'.
SELECT * FROM kna1 INTO TABLE ikna1
  WHERE KUNNR LT '0000100012'.
LOOP AT ikna1.
  WRITE: /45 ikna1-kunnr.
ENDLOOP.
```

Results in the following output:

```
-------------------------------------------------
knal          *knal             wknal         iknal
100002        100002            100002        100002
100003        100003            100003        100003
100004        100004            100004        100004
100005        100005            100005        100005
100006        100006            100006        100006
100007        100007            100007        100007
100008        100008            100008        100008
100009        100009            100009        100009
100010        100010            100010        100010
100011        100011            100011        100011
```

G.6 – **VARY[ING]**

Example

```
DATA:
  prodname(6),
  prodqty(2)  TYPE P,
  prodcost(6) TYPE P DECIMALS 2,
  extended(8) TYPE P DECIMALS 2,
  taxrate(4)  TYPE P DECIMALS 4,
  taxpaid(8)  TYPE P DECIMALS 2,
  BEGIN OF product,
    name1 LIKE prodname, qty1  LIKE prodqty,
    cost1 LIKE prodcost, name2 LIKE prodname,
    qty2  LIKE prodqty, cost2 LIKE prodcost,
    name3 LIKE prodname, qty3  LIKE prodqty,
    cost3 LIKE prodcost, name4 LIKE prodname,
    qty4  LIKE prodqty, cost4 LIKE prodcost,
    name0 LIKE prodname,
  END OF product,
  BEGIN OF tax,
    rate1 LIKE taxrate,
    rate2 LIKE taxrate,
    rate3 LIKE taxrate,
    rate0 LIKE taxrate,
  END OF tax.
product-name1 = 'CMR003'.  product-qty1 = 3.
product-cost1 = '0.23'.    product-name2 = 'CST026'.
product-qty2 = 23.         product-cost2 = '12.66'.
tax-rate1 = '0.08'.        tax-rate2 = '0.0825'.
tax-rate3 = '0.0915'.
CLEAR product-name0.
CLEAR tax-rate0.
DO
  VARYING prodname  FROM product-name1  NEXT product-name2
  VARYING prodqty   FROM product-qty1   NEXT product-qty2
  VARYING prodcost  FROM product-cost1  NEXT product-cost2.
  IF prodname IS INITIAL. EXIT. ENDIF.
```

```
  taxrate = 1.
  SKIP.
  WRITE: /
  'ProdName   Cost   Qty    Extended      Taxrate    Taxpaid'.
  extended = prodcost * prodqty.
  WRITE: / prodname, prodcost, prodqty, extended.
  WHILE NOT taxrate IS INITIAL
    VARY taxrate FROM tax-rate1 NEXT tax-rate2.
    IF taxrate IS INITIAL. EXIT. ENDIF.
    taxpaid = extended * taxrate.
    WRITE: /39 taxrate, 50 taxpaid.
  ENDWHILE.
ENDDO.
```

results in:

```
   ProdName    Cost   Qty    Extended  Taxrate    Taxpaid
CMR003         0.23   3          0.69
                                       0.0800         0.06
                                       0.0825         0.06
                                       0.0915         0.06

ProdName       Cost   Qty    Extended  Taxrate    Taxpaid
CST026         12.66  23       291.18
                                       0.0800        23.29
                                       0.0825        24.02
                                       0.0915        26.64
```

Appendix H - Program Shell ZSPSHEL

Here's an example of an empty program, showing the several blocks it could contain in the order they typically appear. You may start with this shell for any new program to have consistent and convenient code blocks, and remove the blocks you don't use. This shell provides for the use of an external version control system to maintain the code.

```
*$Header: (filled in by version control system) $
*****************************************************************
* (c) (copyright information here)
*****************************************************************
*  Program Name: Z_______ - title
*  Description:
*  Updates Tables:
*  Input Parameters:
*  Output Parameters:
*  Return Codes:
*  Special Logic:
*  Includes:
*******************************************************************
*  MODIFICATION LOG
*******************************************************************
* Date        Rev #   Programmer     Description
* ----------  ------  -------------- ----------------------
*             1.0                    Original Created
*
*******************************************************************
REPORT Z_______ MESSAGE-ID ___ LINE-SIZE ___ LINE-COUNT ___.
*
*******************************************************************
*  TABLES
*
*******************************************************************
*  SELECTION SCREEN (SELECT-OPTIONS & PARAMETERS)
*
*******************************************************************
* TEXT ELEMENTS
*
*Selection Texts
*  parameter name       ' '
*  select-option name   ' '
```

```
*Text Symbols
*   TEXT-001  ''
*   TEXT-002  ''
*...

*Titles and Headers
*   Title              ''
*   List Header        ''
*   Column Header 1  ''
*   Column Header 2  ''
*   Column Header 3  ''
*   Column Header 4  ''
*
*******************************************************************
*  DATA
*
*******************************************************************
*  INITIALIZATION
*
*******************************************************************
*  AT SELECTION-SCREEN (all variations)
*
*******************************************************************
*  AT USER-COMMAND
*
*******************************************************************
*  AT LINE-SELECTION
*
*******************************************************************
*  TOP-OF-PAGE
*
*******************************************************************
*  TOP-OF-PAGE DURING LINE SELECTION
*
*******************************************************************
*  END-OF-PAGE
*
*******************************************************************
*  START-OF-SELECTION
*
*******************************************************************
*  GET & GET LATE events
*
*******************************************************************
*  END-OF-SELECTION
*
*******************************************************************
```

```
*   FORMS
*
*****************************************************************
*$Log:    (filled in by external version control system) $
*
*****************************************************************
* End of Report: Z_______
```

Appendix I - York-Mills Notation
SAP Interactive Command Recording System

Introduction

SAP provides huge capability to interactive users who enter commands at the menus and screens. We found it difficult to discover and remember the sequence of commands to produce a result we wanted, so we resolved to develop a convenient and concise way to record those commands.

The following system (the "York-Mills Notation") satisfies that desire, is easy to learn and record, and can be printed in character and graphics environments. We developed it during the July 1996-1997 ABAP Programming Class at SAP Canada, whose offices are near the York-Mills subway stop on the Yonge line, hence the name.

Here's a simple example of its use, showing how to reach the ABAP Editor from the R/3 Initial Screen:

`/S000 {Tools {ABAP Workbench [ABAP Editor`

Interpret that command sequence as follows:

`/S000`	→ *transaction code of the initial screen*
`{Tools`	→ *press the "Tools" menu item*
`{ABAP Workbench`	→ *press the "ABAP Workbench" menu item*
`[ABAP Editor`	→ *press the "ABAP Editor" button*

and you're there.

Purpose

The purpose of the York-Mills Notation is to record and publish the command sequences needed to accomplish work in R/3. This system is intended as an off-line command record; there's no provision to automate R/3 operations. It includes the ability to package or bundle together frequently-used command sequences; we call those packages macros although they are not expanded.

Delimiters

Delimiter	Command Action
{	Menu item
[	Button
*	Radiobutton
$	Checkbox
<	Field label for data entry
\	Field label for drill-down (Double-click)
/	Transcode
&	Select object (Single-click)
F	Function key (1-12)
!	Shift prefix (Shift-fnc keys)
^	Control key prefix (hot keys and ctrl-fnc key)
@	Alt key prefix (hot keys and alt-fnc keys)
"	Comment to end of line
#	Macro call
()	Grouping
...	Repetition, n ≥ 1 for example `(&field [choose)...`
..	Repetition, n ≥ 0
|	Alternative, for example `(/s000 {tools {abap workbench)` | `/s001`
~	Optional for example `~&keyfield`

Command Action

Always show your starting point such as /S000 for the main menu, or /SE38 for the ABAP Editor.

The command action record may be shown as the name of the control or field involved (such as [Enter) or as the graphic icon for the control and must follow the delimiter with no intervening space. It should be recorded using the same case as the on-screen string to reduce ambiguity. Separate successive commands with spaces or line breaks.

Assign a value to a label if it is required (for example *Maintenance Type = 'One-Step'). The label will be shown without an assigned value if you must enter your own value (such as <Table Name or <Short Text).

Line Breaks

Comments are terminated by line breaks. Don't split a command action between lines. Other than those rules, line breaks are insignificant.

Macros

A macro is a list of commands, recorded with this system, that will be needed more than once and is therefore packaged under its own name. It is called in the command list where the commands it represents are needed. It is simply a shorthand way of recording frequently-used command sequences.

Name: The macro name may be any character string without spaces and starting with a letter.

Definition: The macro definition has a header, a list of commands and a terminator. The header and the terminator are the case-insensitive keywords "Macro" and "Endmacro". The commands are recorded using the notation described herein.

Call: Call or invoke the macro by inserting its name in the command list where the commands it represents are needed, preceded by the macro delimiter (#).

Nesting: Macros may call other macros, but recursion is not allowed. You may nest as deeply as you wish, but deeply-nested macros will be difficult for others to interpret; good practice suggests you limit nesting to no deeper than three levels.

Style Convention

It's good practice to head, foot and comment your command list so you and others can readily know its purpose. Gather commands together in natural groups and use indentation and blank lines to make the command sequences easier to read.

Examples

```
"===========================================================
"Close a runaway SAP Session
"SAP/R3 Release 3.0F
"08/06/96 Dennis Barrett
/SM04  "In another session
&the offending session
[Close
"===========================================================
```

```
"Create a Simple Maintenance Dialog
"SAP/R3 Release 3.0F
"07/29/96 Dennis Barrett

"demonstration of a macro although it's used only once here
Macro Choosefields (&Fieldname [Choose)...[Copy EndMacro

/SE11 "ABAP Data Dictionary
<Tablename   "the table for which you are preparing the dialog
[Change {Environment {Gen. maint. dialog
<Authorization Group
*Maintenance Type = 'One-Step'     "this is for the simple dialog
<Overview Screen = '100'
*Standard Recording routine
"Here choose the fields to include in the dialog
[Fields #Choosefields
...
"=====End of Command List=====================================
"============================================================
"Create a Match Code Object
"SAP/R3 Release 3.0F
"07/29/96  Dennis Barrett
Macro Choosefields (&Fieldname [Choose)...[Copy EndMacro
/SE11 "ABAP Data Dictionary
<Object name    "the name of the Match Code Object
[Create <Short Text <Primary Table
[Tables
[Yes "Save before terminating Editing?
<Development Class [Save [Enter
[Choose Sec. Tab. "defaults to foreign key link
(&Check Field [Choose) [Copy [Copy
[Fields
[Yes "Save before terminating Editing?
[Enter [Choose Fields #Choosefields
[Save [Back [Activate
"Match Code Object is now created and activated.
"=====End of Command List=====================================
"============================================================
"Attach a Match Code ID to a table
"SAP/R3 Release 3.0F
"07/29/96 Dennis Barrett
/SE11 "ABAP Data Dictionary
<Object name    "the name of the Match Code Object
[Change [Matchcode IDs
[Yes "to Create
[Enter <Matchcode ID
```

```
[Enter <Short Text
[Choose Sec. Tab.
(&Tablename)     "Select all the secondary tables
[Enter
[Fields [Yes [Save [Enter
[Choose Fields #Choosefields
[Back [Yes [Activate [Back
"Match code ID is created and attached to tables
"=====End of Command List=====================================
```

About the Author

Dennis Barrett works for SAP America as an Global Support Manager. For three years he was an implementation consultant for SAP. He has programmed and developed database applications for over eighteen years in various programming languages. He received his Certification as an ABAP Programmer from the SAP Partners Academy in 1996 and has been writing ABAP programs ever since. Dennis graduated in Mathematics and Physics from the University of Texas in Austin, and has attended many courses provided by SAP in ABAP programming and in application use and configuration. He was the founding chairperson of the Small and Medium Enterprise Interest Group and the Central-Texas Chapter of the Americas' SAP User Group (ASUG).

Printed in the United States
701800003B